Object-Oriented Analysis and Design with Real Life Applications

(A Trusted Operating System Approach)

Object-Oriented Analysis and Design with Real Life Applications

(A Trusted Operating System Approach)

Dr. Nitish Pathak
Associate Professor,
Department of Information Technology,
Bhagwan Parshuram Institute of Technology (BPIT),
(Aff. To GGSIPU New Delhi)

Dr. Neelam Sharma
Assistant Professor,
Department of Artificial Intelligence & Machine Learning,
Maharaja Agrasen Institute of Technology (MAIT),
(Aff. To GGSIPU New Delhi)

NAMYA PRESS.com

Published by

Namya Press

USA: 10685-B Hazelhurst Dr. # 31013 Houston, Texas USA - 77043
Email: namyapress@gmail.com
Website: https://namyapress.com

Edition	:	First Published in 2022 (Namya Press)
Title	:	Object-Oriented Analysis and Design with Real Life Applications
Authors	:	Dr. Nitish Pathak & Dr. Neelam Sharma
ISBN	:	978-1-956861-27-3
Copyright	:	© Dr. Nitish Pathak & Dr. Neelam Sharma 2022 All Rights Reserved

Acknowledgements

No one walks alone and when one is walking on the journey of life just where you start to thank those that joined you, walked beside you and helped you along the way.

It's with great happiness that we would like to acknowledge many people that get helped us, through the entire difficult, challenging, rewarding, and interesting path towards some sort of Book. without their help and support, none of this work could have been possible.

Through this acknowledgment, we endeavor to thank every person and element, including our guru that has been instrumental in shaping this Research and compiling them to bring it into the form of a book. It also reflects the views of the Guide and Co-Guide highlighted in this book and we remain indebted to them for helping us.

We would like to thank GOD, college staff, HoD, management and the director of our college for their constant support and motivation & for showering their blessings to write such a book on Object-Oriented Analysis & Design with Real Life Applications.

Every single person that we have come across, during this journey has shaped our personality in some manner or the other, a part of which is probably reflected in this book. We wish to thank each of the above members sincerely and wish the best for them always.

Last but not the least, we want to thank our parents, for always supporting us and inspiring us to bring out the best in ourselves, in education, and in everything else. Many thanks. We are grateful to our son, Yash Pathak, and other family members for their unlimited love and strong support in order to write this book.

Dr. Nitish Pathak
Associate Professor,
Department of Information Technology,
Bhagwan Parshuram Institute of Technology (BPIT),
(Aff. To GGSIPU New Delhi)

Dr. Neelam Sharma
Assistant Professor,
Department of Artificial Intelligence & Machine Learning,
Maharaja Agrasen Institute of Technology (MAIT),
(Aff. To GGSIPU New Delhi)

Preface

After the introduction and experience with a series of programming paradigms, object-oriented (OO) software development has become the undisputed mainstay approach to the development of modern software. The reason for this is that the OO paradigm best satisfies demands for the fast development of reliable and maintainable software. Its principle of viewing software as modules modeling real world objects is a natural reflection of both clients' and developers' views of the world and provides a basis for building extendible and reusable libraries of individual objects and whole frameworks that can be relatively efficiently extended or customized to build new applications. The goal of this textbook is to introduce concepts of object-oriented software development and the programming language C++ as a vehicle for their implementation. The novel integration of security engineering, with the Object-oriented model-driven Software expansion approach, is being used in this book.

Its uncompromising reliance on OO principles and its basic simplicity also means that if you want to learn object-oriented programming, there is nothing in C++ to distract you. Whole Computer Science curricula can be based on object oriented software development and C++. The OO principles of all major OO languages in use today have been built on the basis of the C++ experience and are still borrowing principles from it, without ever quite achieving its power and simple elegance, in our opinion.

How this Book Differs from Other Object-oriented Modeling Books

Although there are many excellent texts about object oriented modeling, this is the only one that covers all the topics such as OO concepts, OO application development, forward engineering, reverse engineering, round trip engineering, and programming environment. The book does not assume any previous programming background but the subject matter is not limited to essentials and provides enough depth to allow the reader to write substantial applications and to understand the deeper principles of object oriented modeling. We begin with a presentation of the principles of the object-oriented paradigm, continue with a rather detailed introduction to object-oriented analysis and design, and user interface building tools and components.

This book aims to support the efficiency of object-oriented modeling in a secure software development environment. The conversion of visual UML models into C++ source code helps developers from the analysis phase to the maintenance phase of SDLC. We use object-oriented conceptual modeling techniques to design and develop various applications like E-commerce, Banking, Comparison shopping, Ticketing, Online insurance policy management, product purchase system, etc. In this book, we propose code generation in C++, Class identification, and Modeling for web applications through UML 2.0. Further, we explained the Security Performance flexibility (SPF) model to maintain the balance between security and performance for web applications. The novel integration of security engineering, with the Object-oriented model-driven Software expansion approach, is being used in this book.

In this book, we have described empirical round-trip engineering, object-oriented modularization, source code structuring, and restructuring of a secure software system. Then we locate and understand the most important classes and hierarchies and Message Passing Coupling in classes etc. This book illustrates the abstract, cohesion of classes through forward engineering in C++. In this book, we have presented a SPF) model for secure web applications development. We have also analyzed the integration of security engineering into a model-driven software development approach.

The Scope of this Book and its Intended Audience

Since the book does not make any assumptions about the reader's background, it is suitable for newcomers to object-oriented analysis and design & trusted operating systems. This book is good as a primer and great as a knowledge booster for those already familiar with object-oriented concepts. However, since the scope goes well beyond essentials the book is also of interest to those who already know another programming language and want to learn object-oriented approach. Readers who already know object oriented concepts will also find the book useful because the presentation is quite different from that of other books and includes material not found elsewhere. This includes examples of conventional abstract data types, UML diagrams, use case diagrams, use case descriptions, class diagrams, sequence diagrams, code generation in C++, and the use of the security performance flexibility model of trusted operating systems. Besides, object-oriented approach for secure software development is always worth looking at from a different perspective because there is never an end to learning something new about this rich and flexible object-oriented environment.

Dr. Nitish Pathak
Dr. Neelam Sharma

Contents

1

Trusted Operating System Security

1.1 BACKGROUND

In the last decade, there has been vast growth in the field of networking, sharing of data worldwide. And then comes the most extensively used thing INTERNET have made cyber security a very crucial aspect of work and development. Its matter of concern for both the common users and workers connected all over the world. Despite of lot of works undergoing we are still unable to get something that reliable and silver bullet that it may provide us with complete security for our systems. Being so advanced we still lack the basic potential to create such a system that is capable of stopping viruses and accessing our confidential data from our systems. According to the reports available on Defacement Archives web portal we are very much prone to data insecurity, given by Martin F. Arlitt and Carey L. Williamson, in 2013, see [1]. The security methods developed, worked till yet are implemented in the application layer of the computers which is making our systems more prone to data insecurity. These methods includes encryption using a key i.e. cryptography, using firewalls, access control using authentication, and application layer access control. The most two burning domains are Cryptography and authentication techniques in which max work is being done. Although these are something very difficult to crack but no one knows the dynamic minds making some of probability of data insecurity, was proposed by Przemyslaw Pardyak et al., in 2011 [2]. To some extent using firewalls and application layer access control have helped us but they do have a drawback. These two techniques can help in stopping the attacks using viruses uploaded on internet but fails to protect from internal security issue thus finally making our system vulnerable, was explained by Jose Brustoloni et al. in 2014 [3].

The biggest threat to our application layer is viruses and Trojan Horses. Once these two enters in our system they have the potential to access and even modify each and every data present on the system. Now these days, to overcome the threats operating system application layer and the network entry points is used to implement the security measures. Although no preventive measures are used inside the kernel of Operating Systems. It is believed that security measures in kernel are much more effective than the application layer. In fact, after lot of work such operating systems have been developed which have much more mechanisms inside the OS kernel providing us very good level of security thus securing our systems, was proposed by Hao-hua Chu et al. in 2015 [4]. A shocking stuff to know is that the technologies developed for OS kernel are not new and were developed very long before around in 1970-80. The cost and the complexity led to its failure of reaching the wholesome consumers and are widely used only in organizations which need high security and can't compromise with it.

In 2013, Dawson R. Engler et al. suggested that the Security is not something expected not only by big organizations but also by common consumers so now concerns are being there on this and many vendors are trying hard to fix the issue. The companies which came up with some promising operating systems with security features are Argus-Systems Group, HP, and Sun Microsystems. Open Source OS are also providing well secured kernel having excellent security features and commonly known as secure systems, operating systems [5]. National Security Agency has released the most secured and that too open source operating system called as SEL linux. Proper definitions of secure system vary from organization to organization. In this book Trusted Operating System term will be used in order to refer such systems.

1.2 TRUSTED OPERATING SYSTEM SECURITY

Before moving into the problem, we are dealing, first we will be talking about the basic principles of Secure Operating Systems. As was mentioned previously, the term Trusted OS is interpreted differently and vary from one company to another software company. During system programming, company develops the system software according to the requirement of end users. But there are some important features in all Trusted Operating Systems. They are as follows Least Privilege, Mandatory Access Control (MAC), Discretionary Access Control (DAC) and auditing, was designed in 2008, by Sylvia Osborn et al. [6]. In this section, we will be discussing about some history of Trusted Operating Systems.

1.2.1 Trusted Evaluation Criteria for Operating Systems

We have been working for better security methods since 1984s. We were not sure about the appropriate requirements what operating system must have. In this regard, the trusted System judgment Criteria was defined by US Department of Defense in 1983. After these basic guidelines were set and operating system were programmed and evaluated accordingly. Under the joint effort of two giants Europcan and North American Nations, Common Criteria was introduced which became the essential part of these evaluations, was introduced in 1984, by Soloway, E. and K. Ehrlich [7]. Despite of being most modern Common Criteria, mostly recognized evaluation criteria is still TCSEC. What actually TCSEC did was to combine all the ideas into one but the best part was developing a rating system. The ratings system determines the actual strength of security system in our operating system. Ratings or strength is known to be of 4 classes A, B, C and D resp. and A with max security and D the least one. For more additional information check or for a complete knowledge see [8].

Class D has the least security measures build into OS. It will be shocking to know that MS-DOS and Windows 95/98/Me have got class D ratings. Class C operating system provides Discretionary Security Protection or Discretionary Access Control (DAC). These operating systems are even capable of auditing. Brief description of both the features will be there in sections 1.2.2 and 1.2.5 respectively. UNIX, Linux, and Windows NT which are even used today are rated as class C operating systems. Class B systems provide Mandatory Security Protection or Mandatory Access Control (MAC) and all other features of class C operating systems, was explained in [8][9], in 2006, 2005 respectively. Brief description about MAC is given in section 1.2.3. Most common examples of class B operating systems are Hewlett-Packard's Virtual Vault, SELinux, Argus System's Pitbull and Trusted Solaris. Sun Microsystems developed trusted solaris as much more secure system to maintain the security aspects in applications, was proposed in 2009, by Andrea De Lucia et al. [36]. Despite of fulfilling almost all the requirements of class B system, it still fails to fulfill guidelines of TCSEC and is not given a class B rating. Class A operating systems are one which are widely used in military domains and provides tight security features necessary to secure confidential information [58]. Commercial vendors do not provide such operating system in the market for common use.

1.2.2 Discretionary Access Control (DAC) Security Technique in OS

In 2005, Kölling et al. explained that the security method is a key feature that helps the user to get information about who is accessing the objects and what will be the procedure to do that. For example, using this DAC mechanism, administrator may share some of its files with other user and providing access such as read, write and modify the file , was made available in [10]. The best examples around us of DAC are UNIX and LINUX OS. Every file this OS is provided with set bits that tell us read, write and execute permissions. Each user has the rights to change the permissions of each file. Administrators have rights to grant permissions to even themselves, another user and anyone in the group. It is significant to note that DAC permits computer users to change permissions for even themselves. These mechanisms are like grant and revoke mechanism. This features helps in preventing human error, was introduced in 1998, by R.W. Selby et al. [37]. For example, this mechanism can stop from deleting a file by mistake which may be very important for the system. Most widely used security technique in OS is DAC. Unfortunately, we cannot just rely on DAC and expect high security since it lacks the ability to protect our confidential data and wholeness of data thus destroying wholeness of our data. Moreover, DAC is inefficient in protecting our data from authorized users from deleting or modifying data, was proposed in 2011, by J.M. Fernandes et al. [38].

1.2.3 Mandatory Access Control (MAC) Security Technique in OS

In 2002, H.C. Betty et al. explained that MAC can be most excellent explained as security mechanism that is not in the hand of the user rather it is controlled totally by the operating system and the guidelines set by the system administrator. Thus, MAC controls the flow of data much better than DAC. The reason for better security can be best explained using their implementation and modeling. MAC is the anchor, providing us the desired security in trusted operating systems [39].

Security labels are used to implement diverse MAC code which ultimately confines and deals with many of the security issues. Despite of weakness in DAC, still it is irreplaceable and both DAC and MAC are implemented together in a system allowing some extent of security discretion for computer users. MAC main task is to limit how the user will modify or change the security permissions residing in the system. In addition to this, permissions with the system administrators are also taken into consideration by MAC security guidelines. This is not as same in DAC in which root processes are allowed to bypass, which can also be found in literature [40]-[43].

1.2.4 Least Privilege Security Technique in OS

According to this access approach, only those privileges or permissions must be given to users and processes required to complete certain task. It is very difficult to decide the minimum permissions one operating system must provide to any user or process. However, giving access more than required generally leads to security threats. There are many techniques least privilege can be implemented in our operating system, introduced in 2006, by Marco Brambilla et.al. [44]. One way is to use proper security labels in MAC mechanism. However, the best and well accepted way to apply least privilege is via mechanisms like to capabilities. The permissions in computer system tell us that what all permissions are given to the process and what all actions are allowed to take during executions. In the computer system, each and every process holds list of permissions and capabilities. These capabilities store the list of actions or permissions which a process will be allowed for the period of its lifetime span. With the above security technique, we can make our system well secure using least privilege.

Anyways a system with a user with all permission's can be very dangerous. It makes the attacker do whatever he wishes to do because he has access to the whole system and even the root account. Least privilege along with DAC and MAC helps lot in making the system more secure.

1.2.5 Auditing Security Technique in OS

As we have revealed previously in Section 1.2.1. Another significant aspect of security is Auditing in Trusted Operating Systems. It actually records all the security operations and transactions in a system. Logging information actually helps us in getting the source of attack or does the system can expect any other attack in future, introduced by Guindon, R. and B. Curtis, 2009 [15]. Most systems available in the market are programmed in such a way to store basic logging information but fail to do the high grained logging feature which is mainly required in system with high security. Systems currently available cannot store data about the user that which file he is accessing which is very credential data in retrieving security holes in our programs, available in the literature [16]. In 2001, S. Tanenbaum explained that the secure Operating Systems or TOS are well programmed to do the fine-grained auditing as well capable of restricting user to modify the stored or logged information. Another security measure taken is to encrypt the file containing the logged information because without it any attacker can access it or hide his actions. OS must be well programmed that it should effectively determine what all actions need to be logged so that better analysis can be done in order to have trespass detection. For this book, auditing is a concept in which some selected set of actions of user are being logged and measures are taken to protect the log file against modification [17].

1.2.6 Structural Design of Trusted Operating Systems

The essential structural design of Trusted Operating Systems is shown in Fig. 1.1. Just for a reminder to the readers; architecture is just a concept although implementation can be done in a lot of ways. The architecture of traditional operating systems is given in Fig. 1.1(a) System call interface helps the application and middleware interface to communicate with the Operating System.

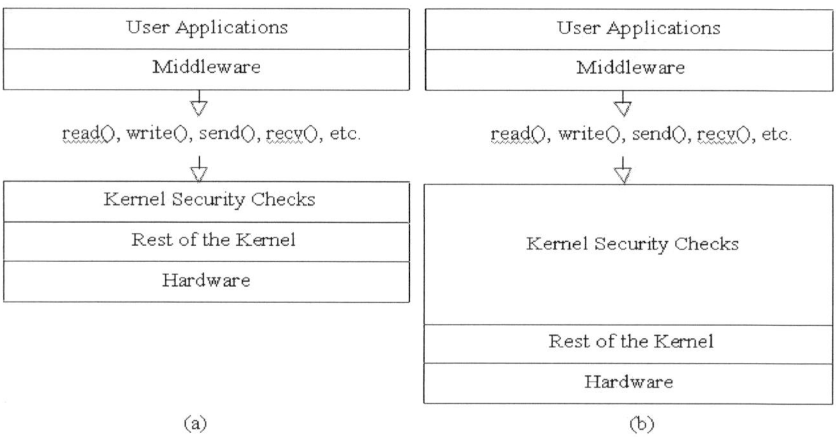

(a) Standard Operating Systems. **(b) Trusted Operating Systems.**

Fig. 1.1: Design of trusted operating systems and ordinary operating systems adopted from [1]

In most commercial operating system, we cannot expect much of security. The security in kernel is just fair enough to implement DAC. This is shown in Fig. 1. 1 (a), illustrating thin or slim security layer of operating systems kernel security checks, was made available in

[8][18]. Now in order to provide higher security lot of security checks are there in kernel of Trusted Operating Systems. 1.1(b) clearly depicts the thicker layer of kernel security checks. What all security measures are being taken in the kernel security check depends all on implementation and modeling. This layer may include least privilege, auditing, MAC, DAC and much more, available in the literature [36]. But the disadvantage of having extra security check is that whenever user tries to do any useful work it need to undergo all the checks thus deteriorating the system overall performance.

1.3 CHALLENGES IN TRUSTED OPERATING SYSTEM SECURITY

As we have told earlier the main domain of Trusted Operating Systems is Military and lots of efforts are being made in order to make them suitable for commercial organizations. We have a lot more discussion regarding this in Chapter- 2. Performance is one of domain of trusted operating system which has not been worked yet. As the architecture, Fig. 1.1(b) depicts the thicker layer of kernel security thus slowing down the overall performance. Thus, low performance comparatively to commercial operating system makes it less popular among the common users, introduced by Ford, J.K. et.al , 2007[17]. More the security measures in kernel of operating system more the system will become slower.

In 2011, Pedro Valderas and Vicente Pelechano suggested that in some areas apart from military, high security is not that much necessary. Best example around is web servers. A typical web transaction goes under fourteen operations. Out of these 14 operations, 09 operations or system calls or member functions or member methods must pass or go through extra security checks for accessing the data like from the disk and sending it on a network socket. Web server deals with sole public data and public information, introduced [19]. Since majority of data is public on any web server, task of checking it during read from disk is something useless because this data is already readable by each and every user using internet. The real task of security comes when it comes to writing access. For any web server integrity is the main issue rather than its confidentiality.

In 2011, Pedro Valderas et al. state that during various number of security checks even the multimedia and video streaming services are also being compromised. On a commercial operating system whenever a video is played each frame is picked and read from the disk and displayed one by one on the computer or on the other display screen. In the case of video streaming frame is sent out onto network using sockets. This all functioning is possible through socket programming. Security of TOS also affects the quality of multimedia and video streaming services. But in case of trusted operating systems for every read from disk system have to do repeated checks. Due to repeated security checks, the frame rate and quality of video streaming will be decreased. The result will be worst when the running system is heavily loaded with lot of programs. In this case when system is very loaded with multiple processing and multiple applications are in running mode, definitely the streaming will be very slow and the quality of video may be very poor. This isn't true for Video On demand server because they keep in mind about the security checks so they care about the video quality rather than read accesses to the server. This selection of operations and system calls will be different from one application to another. So for the selection of these necessary and un-necessary operations pattern mining is required, introduced by Andrian Marcus and Denys Poshyvanyk, 2008 [21].

1.4 SOLUTION METHODOLOGY IN TRUSTED OPERATING SYSTEM SECURITY

As we had stated that there are many types of workloads that are continuously being checked by the security mechanisms of kernel in which many of them are very much useless or undesired in a Trusted Operating System. Primary concern of these workloads is quality as well as integrity of data rather than security of certain operations since the data they consist can be public. This let us conclude that by disabling security measures of some parts of OS performance can be increased, was introduced in [8][22]. Performance can be improved by providing administrators the permissions to balance security and performance needs.

This book proposes a concept of SPF (security performance flexibility) in order to gain improved performance and speed for particular system workloads for Trusted Operating Systems, available in the literature [23]. This particular proposed model allows system administrators to skip or disable some unnecessary security checks in these operating systems through which they can effectively balance their performance needs without compromising the security of the system. Problem Solution Methodology is illustrated in Fig. 1.2. The modeling and implementation idea behind the SPF structure is shown in Fig. 1.2. Thus providing administrator's option of disabling security checks for some useless system processes or undesired process, was made available in [8] [24].

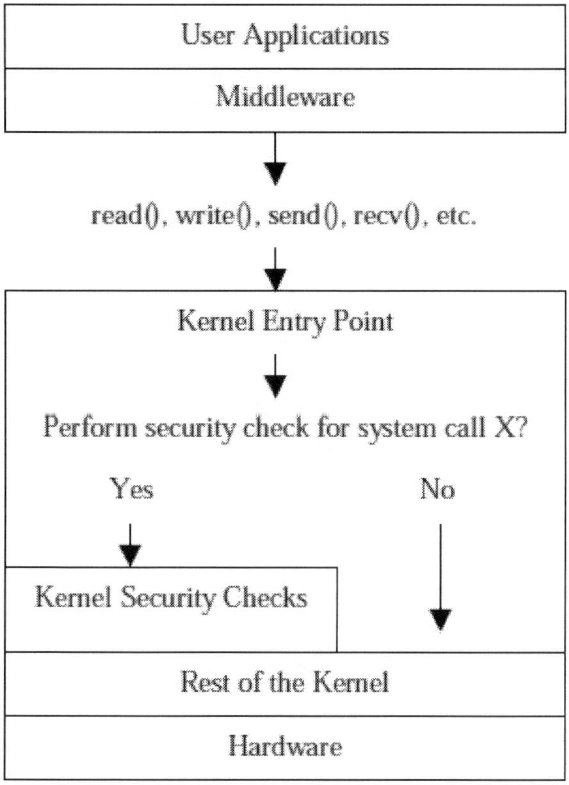

Fig. 1.2: Security performance flexibility framework architecture adopted from [2]

By skipping these processes or system calls the performance of system will be improved. These SPF based improved version of secure systems can be used for desired web application, for maintaining the desire level of security concern in particular application, which can also be found in literature [8]. The performance of the systems surely increased. For example, admin of the system can disable all the read checks in web server because they are actually useless which finally increases throughput of the web server [25]. There are many levels at which we can implement SPF framework in Trusted Operating System. For evaluations in this, these two different levels are being considered.

1. System based SPF representation for web applications.
2. Process based SPF for web applications.

Disabling trusted security checks for certain set of operations of the system is done via System-SPF. For example, System SPF can disable all the read security checks from the system and thus as a whole we are improving the security performance up the desired level in system, which can also be found in literature [8],[26]-[28]. This System-SPF is useful in web a server which generally deals with pure public information thus increasing web server throughput. However, one disadvantage of System based SPF is that it does not provide very well grained facility for administrator to manage the system. As an example, whenever a system admin wants to disable just some number of applications security checks but for not all the applications installed, P-SPF or Process-SPF provides the system admin enough permission to disable or skip the security checks in particular applications or user processes / system process. For example, administrator may uncheck or disable the security check on read operations in any video player thus whenever a video will be played security read check will be skipped and finally it will increase the frame per second and overall quality of the video streaming, was made available in[29] [30].

In 1996, Rosson et al. proposed that O-SPF or Object based SPF gives system admin with a special fine-grained procedure to manage or control their system. This object based SPF model is developed according to the requirement of system. Disabling security checks on individual objects in the particular system is well done by Object-SPF. Objects can be anything such as files; network connections etc. Let considers our previous example of video player, which can also be found in literature [31]. Now Let us consider some system in which we have both public as well as private videos. We may want to skip or disable to security checks of public videos but not the private videos. Object based SPF model allows system admin to differentiate between files in which security checks are been skipped. In the same way, this allows system admin to skip the security measures for network connections like message queues or sockets, introduced by Wasserman et al., in 1994[118].

1.5 SOLUTION EVALUATION

For various reasons Security Enhanced version of Linux was chosen. First, it is one of the popular operating systems around the world. Thus, for studying SPF Framework Linux is best suitable because it provides us the performance which is required now these days. Secondly, SELinux security is combination of different mechanisms which are based on Multi-Level Security. The selection of this multi-level security will vary from one web application to another, explained by Hoc et al., in 2010, which can also be found in literature [27]. The security SELinux provides cast the work done on security measures in order to secure the system in most effective way. Added advantage of choosing SELinux is being open source thus allowing modification and change as per your requirement. Just because of privacy and confidentiality in Trusted Operating Systems, the source code of any software company, business and armed forces will not be available for normal user. So obtaining such source code in specific language is not as easy as we think. The privacy and security implementation for any system will vary from one Development Company to another [32].

The available open source code of SELinux is its most important advantage; this available source code is also one of the main weaknesses of SELinux. Normally, the source code of such operating systems is not deployed to the open public use, in view of the fact that security holes can easily be fetched from the code itself. SE Linux is mainly being designed and developed as a work project for the users of Linux community, available in the literature [33][34].

In 2015, Ibéria Medeiros et al. reported that Security enhanced version of Linux is persistently changing, as various innovative thoughts, features and additional aspect of security are discovered and implemented through specific system programming language into the system kernel. Even with above mentioned problem, Security enhanced version of Linux is still a high-quality operating system to assess the SPF structure for web applications [35]. In order to assess and evaluates the capability for SPF, the subsequent criteria will be considered for the same:

1) Performance Improvement

The main objective of the SPF support is to observe how greatly performance improvement and up gradation can be achieved, if any. It is quite probable that the SPF overhead may possibly override the SE Linux security overhead, thus the performance cannot be achieved and increased, as desired by the user.

2) Complexity of System Administration

These secure Systems are more complex for computer administrators to handle and manage. As we know that conventional operating systems can be managed by system admin easily. Such secure Systems require much more extra effort and time to setup the desired security policy on the part of administrator. The implementation of security policies, as per the requirement of user, is very complex in such systems. Just because of these reasons, these systems are not so much popular in commercial sector, in 2005; it was suggested by Rist et al. [28]. It is necessary that the system controlling and management of such systems should be so simple. Otherwise there will be enough burdens on system administrators.

1.6 BOOK CONTRIBUTIONS

The main contributions of this work are given below:

1. The original proposal that not the all and entire security in Trusted Operating System is required.
2. The fresh initiative that skipping irrelevant security checks in such System may improve the performance of system in all respect.
3. The SPF model can be designed and implemented at various levels like system based and process based implementation is possible. This selection will be under consideration according to system's need.
4. This work allows system administrators to configure and arrange such systems dynamically as quickly as the system executes.

This work creates an experiential base to showcase the effectiveness and efficiency of SPF based object oriented modeling mechanism in software development, with primary focus on security in applications. To retain the balance among security and web application performance, we propose amalgamation or combination of security based forward engineering and round-trip engineering. This round-trip engineering in software development is used to maintain synchronization between source code and software design, in 2004 by Lange et al. [29].

This work proposes SPF based secure software analysis, SPF based secure software design, SPF based secure software development and roundtrip engineering based maintenance approach. This book also demonstrates the designing of real-life case studies. We will consider some real life case studies for web applications in chapter 4 for the same proposed scheme. To maintain the legacy of source code, this work also proposes that round trip engineering is necessary for real life applications. Therefore, one should not only focus on the program development but also on system maintenance. Because software and web application do not have wear out phase. Application development is short term process but maintenance, reverse engineering, reengineering and restructuring of source code is long term process and much time consuming, available in the literature [30], by Rosson et al., in 1997. Source code listing and Restructuring of program should be done in such a way that it completely avoids the overhead of reverse engineering of a system again and again because of modifications being made to its code over time. The mining restrictions imposed by proposed information securing system, includes complex processing steps, so it becomes a necessity to visualize and design its model accordingly before the actual implementation is done.

In this book, we describe experiential round trip engineering, object oriented modularization, source code structuring and restructuring of secure software system. We analyze the quality of the system in all and have gained an outline of it in terms of its size, structure and complexity. Then we locate and understand the most important classes and hierarchies and Message Passing Coupling in classes etc. This book illustrates the abstract cohesion of classes through forward engineering in C++. In this work, we present a SPF based approach for web applications and the integration of security engineering into a model-driven software development.

1.7 ORGANIZATION OF THE BOOK

The outline of the book is as follows:

- **Chapter 1:** Chapter one presents the background of trusted operating system security, list of problem statements with description, work carried out against problem statement, brief discussion about the solution evaluation for web applications and book contributions etc.
- **Chapter 2:** Chapter two of this book deals with the literature review and related work, quality of service and security performance flexibility for web applications, mining software data through source code and at the end of this chapter summary of literature review and related work.
- **Chapter 3:** Chapter three of this book presents problem description and solution methodology, secure operations supported by TOS, SPF design assumptions, security performance flexibility architecture and design overview and core concepts of object-oriented development for SPF based development of secure web application.
- **Chapter 4:** We have addressed the implementation of object oriented secure round trip engineering for web applications, in this chapter; we have considered real life case studies and web applications. Chapter four of this book presents the mining based experimental round trip engineering, object oriented modularization, source code structuring, restructuring of secure software system and SPF based round trip engineering for secure web applications.

- **Chapter 5:** Chapter five of this book presents result and performance evaluations of proposed SPF framework for the development of secure web applications. The results of SPF model for web application also has been given in this chapter.
- **Chapter 6:** Finally the chapter six, in this book, has discussed the conclusions of the book and possibilities of future work that can be undertaken and investigated.

2

Operating System Security Policies

There is a negligible amount of work that can actually be undertaken to SPF model. Still, the SPF frame is related to innovative ideas from three distinct areas of work. The first area of work is related to security. The other two work areas are flexibility of operating system and Quality of Service in systems. A lot of quality work has been worked for quite long time in Operating system flexibility; these matters will be further covered in sub Section 2.2. A light summary of Quality of Service in systems and Quality of Security will be shared in Section 2.3 and after this its relation with security will be discussed in precise manner. A brief review of these interrelated works and their relation to SPF approach will be explained in sub Section 2.4.

2.1 INTRODUCTION OF OPERATING SYSTEM SECURITY

Section 1.2 of this book gave a layman summary of conventional Trusted Operating Systems performance, concepts and execution. This ongoing section will cover a number of newest works in operating systems security. Further, Security policy flexibility related to SELinux and other secure systems will be explained in Section 2.1.1. At last, a detailed review of the latest security efforts collectively and their relation to SPF framework will be explained in Section 2.1.4. There are two major dimensions in operating system security that should be pointed out. The ideas are:

1) Making the system administration work easier to maintain web security.

As it has been revealed previously, system administration work of secure TOS is certainly far additional difficult as compare to that of conventional operating systems. Latest study and work efforts have contributed at ways to make system administration work easier in such operating systems. The system administration work should be easier to maintain web security, was proposed in 2001, by Anat Eyal et al. [40].

2) Adding security features to facilitate additional security policies to be imposed in web applications.

In conventional Systems, security and data protection policies are quite inadequate and incomplete to those that can actually fit the security model implemented in the system's kernel, was made available in [45].The person who reads should take notice that the security policies and process discussed and explained in this work section still pursue the identical security framework and structure as described and explained in Section 1.2.6. The structure of these OS is little bit different from traditional OS, so, the programming efforts and implementations of such systems are also different. Very good system programming is required on part of software developers. Without good system programming skills, the implementation of such systems for web applications will not be possible. Still, the security policies and security approaches are limited and still operating systems based and implemented inside the kernel, introduced by Peter Desnoyers et al., in 2012 [46].

2.1.1 Support for Flexible Security Policies

Although the latest work explained previously implement so many security policies at kernel level as well as application layer level to protect the systems and web application, but after this these efforts towards security are still relatively incomplete, imperfect and limited. The implementation of security policy at kernel level is still very limited. In order to suggest a much needed security, operating system structural design must be developed to grasp variable security policies.

The essential structural design of DTOS and Flask is revealed in Fig. 2.1. Unlike the conventional methods of adding more and more security layers at the kernel level, we are suggesting two supplementary or extra subsystems in this structural design. The responsibility of object manager in this model is to call the security server each time whenever a system user tries to access an object. Particular security server confirms the security pattern and informs the respective object manager if permission for requested operation is granted or denied. In this design approach we have to notice that the security server is not at all the component of the kernel. It is a different and separate part that can be called as per requirement, by the kernel. These different modules of the security server also can be altered or changed, was proposed in 2005 by Ilka Philippow et al. [47]. This is the main reason that's why DTOS as well as Flask both are built and implemented upon kernels.

Almost security policies can be implemented if we consider security server as a separate part or separate module. These separate modules can be easily modified as per web applications requirement. This is not hard and fast that security need always will be unchanging or statically placed within the kernel, was made available in [48].These security needs and implementation will vary from one real life application to another. The implementations of web application for maintain the security is very much subjective in nature. The security requirement for the same will depend upon the user's needs. The object manager all the time calls the security server for checking the granted permissions. If security server grants permission for particular operations, then ok, operation or specific system call will be passed to kernel layer for execution. If permission is not granted, requested operations will not execute. The security features and security policies can be altered, as per need, dynamically as the system is in execution phase. Security layer will execute the security checks according to new altered security policies; it was suggested in 2002, by Elbaum et al. [49].

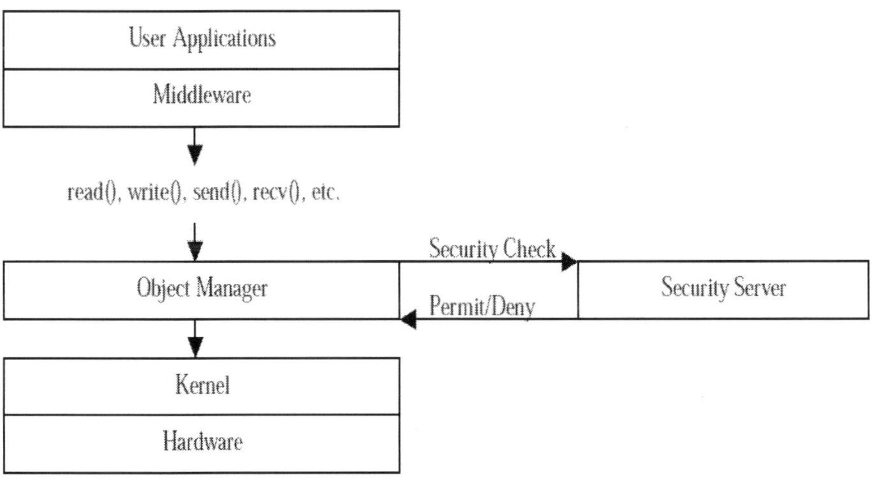

Fig. 2.1: Flexible security policy architecture adopted from [8]

In order to boost performance in DTOS as well as in Flask, security policy caching was recommended as a means to strengthen performance. The mainly referenced security checks are stored in a software-implemented cache, located in the module of object manager; it was suggested in [8]. This can surely increase the performance of systems. It might be predominantly helpful and valuable for many applications such as a web servers, search engines and Video on Demand servers etc. By caching recent security policies in the object manager, few parts of the security check can be ignored [50].In this case, if security policies changed by authentic user, new security policies will be implemented with immediate effects.

2.1.2 Development of Security Performance Flexibility for Web Applications

As stated all over this section, it seems the future of security in operating systems is dependent on two major factors. These two factors or themes are the development of TOS with very easier system administration work and system programming methods to apply a massive variety of security policies. At the time of implementation, these massive varieties of security policies will vary from one application to another. The selection of these securities for a particular web application is very much subjective in nature, was introduced by David Carmel et al. in 2011, [51]. Such operating systems facilitate us to make simpler security source code. Flexible security policies are such characteristic that may proves to be helpful for large number of companies and organizations. On the other hand, this observation will show that flexibility in security is not at all being useful. Efficiency, effectiveness and Performance, of security features should also be considered as a flexible constraint [144]-[146].

This approach proposes that enhancement of SPF in operating systems may turn into extra significant in upcoming Trusted Systems. In 2009, Thiago Jabur Bittar et al. introduced that through this approach; upcoming trusted system for web applications will be more and more refined and will be more qualitative in nature. It will be one of the most important aspects because latest work efforts signify the performance hole or performance gap among ordinary operating systems and trusted operating systems [52]. Due to balance between security and performance, in near future these Operating Systems will become more popular.

2.2 DESIGNING OF OPERATING SYSTEM FLEXIBILITY

In 2006, Luiz A. Ricci et al. introduced that the designing and implementation of security at kernel level is not very new concept. Even implementing flexibility for respective web application into operating systems is not a very latest and novel concept. It has been considered for relatively and reasonably a long period of time. So many eminent authors and work scholars have worked in the same domain. But due to cost, complex implementation, these systems are still not very much popular in commercial sector. Still very much scope is pending in the implementation of such systems [53].

2.2.1 Flexibility in Operating System and Security Performance Flexibility model

In particular, flexibility allows that the operating system to be altered and modified relatively easily. To improve the performance in all respect, the modifications in the system will be carried out by system administrator. Thus, performance improvements in an operating system can be accommodated and incorporated more easily. Flexibility can promote TOS by allowing security checks to be skipped if they are considered meaningless and pointless. By skipping security checks, performance of the system enhances in the system and ultimately can be gained [8][54]. Which operations are necessary and unnecessary, the answer of this question will depend upon the nature of system as well as the requirement of user. This pattern mining can be identified during the process of requirement elicitation and discovery. After this software requirement specification (SRS), software design document (SDD) finally implementation of SPF will be possible. SRS will be done by system analyst; SDD will be designed by software designer. Implementation of SPF frame work for specific web application will be possible through coder or software developer. During the coding, coder will apply the constraints i.e. which operations are going to be implemented and which operations are not going to be implemented for specific web application [55].

2.3 WEB BASED QUALITY OF SERVICE (QOS) AND QUALITY OF SECURITY

The SPF structure is furthermore related to the idea of Quality of Service. QoS is one of the mainly worked topics in information technology and engineering today. Services are available but not qualitative in nature, such services are useless and not desirable in real life application. For example, if any application or software providing all functional requirement i.e. services, but that particular system is not providing the non functional requirements i.e. quality, once again system will not serve our purpose. In other words we can say that every system should cover all functional as well as non functional requirements with perfection. Quality constraints are the major concern for any software. The subject of QoS and its implementation methods can go relatively in depth to depth, was reported by Keng Siau et. al in 2004 [56]. For that reason, merely the basic ideas will be explained in Section 2.3.1. After the review of QoS, its relation to security will be explained in Section 2.3.2.QoS is also one of the most important aspect for real life web applications. To maintain the high quality in web application, the concept of QoS should be considered for further processing.

2.3.1 Required Quality of Service (QoS) for web applications

For the design and development of secure web application, Quality of Service in any application is must. Every web application should provide the high quality services to the end users; it was suggested by Thiago Jabur Bittar et al. in 2009 [52]. Services are required from CPU, Memory, printer, network services and bandwidth etc. for execution. Therefore, QoS is essential for services to match and maintain the resources it needs. So we can say that qualities of services are must for any system. Without it performance of any system cannot be achieved. For example, a service may want to assurance performance at a desired rate or execution or completion of process within the specified amount of time. System is high qualitative, if it provides desired services within the specific time.

The concept of QoS is very much popular for applications like images, pictures, videos and multimedia applications etc. In multimedia applications QoS are very much needed. Let's take an example related to a MPEG video, clip or movie .in this multimedia application playback requires the specific framing and streaming of video frames and bits. These frames to be played at definite intervals apart from one frame to another. The equal interval is very important for the display of frames. If frame rate and interval is up to the mark than the quality of video or movie will be high qualitative.

Let's take an example, if a MPEG video has a frame rate of 30 frames/ second, in this example, each frame should be played roughly 33.3 milliseconds of difference between to video frames. MPEG video should play smoothly, the quality of framing and streaming should be up to the mark. In most of the cases, operating systems does not provide the support for high quality videos in terms of streaming and framing. In ordinary OS there is no fundamental method that can promise the quality of services. In system programming, normally high priority is given to system processes not the user process. It means system process always having the high priority over the user priorities. In case of video on demand, MPEG video frames have lower priority.

Due to limited framing and live streaming, the overall quality of video will not be up to the mark [147]-[149]. If the same is very loaded then the quality of same video will be poorer. When frames are played late, the video will subsequently look irregular as a result low performance.

In this problem, SPF model for this application can serve our performance to achieve high quality video. Although QoS are normally associated with multimedia applications, video on demands and web servers. But the scope of QoS is not limited, it can also be implemented and linked with common applications as well, was made available in [53] [54].these quality of services will be subjective in nature and will vary from one application to another.

2.3.2 Relationship between Quality of Security, Quality of Service and Security Performance Flexibility

Section 2.3 details about how the generic applications and services may need resource guarantees therefore they can complete a desired task with assured security. From this point of view, SPF model is extremely related to QoS. TOS security methods may decrease and degrade the overall performance requirements for any web application, was made available in [55]. The SPF structure will give system administrators the authority to disable or skip the security checks therefore their security performance requirements can be properly met. The development of software systems usually includes their specification, design, implementation, testing, deployment and maintenance phases. The general trend is that software companies often tend to obviate some of the steps (e.g. requirements analysis, specification, design and testing) out of the software development process to release the software as quick as possible due to the strict deadlines, was introduced in 2009 by Per Runeson et al. [57].

2.3.3 Mining Software Data through Source Code

The mining of operations can be identified during the requirement elicitation phase of software development life cycle. After identifying the system call patterns through requirement engineering, implementation of SPF model for web application is possible. The pattern of system call and operations in a particular application can be identified from the existing source code of open source software. This source code can be modified according to our latest requirement. Through the source code in specific programming language, Complexity measures can be identify like information flow between module, McCabe complexity, lines of code, Halstead complexity, data matrices, etc. have been widely explored [58]. The reference [59] validated the effectiveness of the object-oriented design measures. Object oriented measures are like no. of classes, no. of data members in a class, no. of member methods within the class, no. of classes inherited, no. of super classes and no. of sub classes etc.

The most important objective of software development process is to deliver best possible Software, i.e., software with high quality at low cost & efficiency and deployed within the specific time duration. Best possible software means covering all the functional and non functional requirements in all respect. These functional requirements, we will represent through use case diagram and with the use case description. In upcoming chapters, we will cover this use case driven approach for the development of secure web application. In order to achieve this best possible software, programmers normally reuse the existing libraries, rather than implementing similar code products right from the scratch, was reported by Roy et al. in 2011 [62]. Rapid application development is required for the same purpose. Rapid development is possible through the concept of reusability. While reusing the same libraries, software developer/coders usually face numerous changes such as several existing libraries are not correctly formulated and various libraries contains huge number of program interfaces through which libraries get exposed to their functionality in whole.

These problems direct to certain problems that influence in producing finest and best possible software. The development of finest and best possible software is not as easy as we think. The problems like reuse of existing libraries often lead to consumption of more and more time, lack of knowledge on re-usage of program interfaces and also we can't generate effective and efficient test inputs during white box testing. As we know the white box testing is structural testing where we mainly concentrate on the source code of the system [60].

Data mining in e-commerce involves the use of sophisticated data analysis tools to discover relevant web log data sets. This approach can be performed on web log data represented in quantitative, textual, or multimedia forms [61]. Data Mining can be described as: enterprises established business objectives to explore and analyze a large number of organization data and to reveal hidden information. It mainly has four kinds of technologies: statistics, machine learning method, the neural network method and the database. E-commerce is an easy and comfort way for shopping goods or products in which with your own interest you can do shopping. There are many products available online for which you can do shopping which is one of the wonderful experiences. Traditionally there was only one way of shopping, to go to near market or a mall and buy things with cash. The simple structure was pay cash and get what is required. With the evolution of computers and internet, a new virtual world emerged and people started their buying and selling in online shopping stores. That way got popular within no time and business organizations started to make online websites so that people can buy through them. Earlier it was observed people do online shopping due to heavy discount but in reality due to comfort, more brands and variety, people are showing more interest in shopping via online [62].

Organizations rely on digital data which is increasing exponentially these days. Digital data can be efficiently used to discover useful patterns and relationships that can progress business and this can be possible with data mining techniques. Data mining is a practice of inventing novel and significant associations, patterns and trends to scrutinize enormous quantity of data. There are number of sectors like healthcare, forensic science, fraud analysis, banking, marketing, sports, telecommunication, bio medical etc. that are using it on a regular basis. The reason of popularity of this technology in many businesses is because it allows them to learn more about their customers and make smart marketing decisions, was suggested by Baresi L. et al. in 2000 [63]. Healthcare facilities have at their disposal enormous quantity of data which on systematic analysis can lead to more disciplined decision making. Thus, the challenge is to dig out significant knowledge from this data and act upon it in a timely fashion.

Data mining techniques and algorithms are becoming a common field because there is an urgent need for identifying novel and valuable knowledge in enormous data produced exponentially in all sectors for decision making. Various techniques like classification, clustering, association rules etc. are used for discovering knowledge from vast amounts of data. In the present paper, child immunization data has been selected for the purpose of classification. The classifiers partition a dataset on the basis of classes of its patterns and predict the class of an unknown pattern, called test pattern [64]. "Getting and identifying opinion" is identical to get the details about what exactly one thinks about a particular object. Today, this is known concept as opinion mining and it plays very important role in the field of businesses where decision making is critical and vital. From few decades, the internet became a pervasive tool in day to day life, people were asking their friends and families to help them in taking any decisions such as which is the best place to enjoy the holidays or what is best option to purchase online, etc. Because of the source of information, the internet provides so many details about products as their specifications and other details related to reviews, comments specified by other users who have already purchased a product or experienced the services of related companies, was reported by P. Epstein et al. in 1999 [65]. Even, business organizations always curious about consumer's trends, comment on their products and likes-dislikes of their organization's expressed by consumers. This data and its analysis always help to move with proper decisions in business for its further growth. Mining of operations are the major concern for SPF model. Data mining methods on code examples that are gathered and recognize common patterns, it was suggested in [66]-[68]. The regular patterns show accurate usage of program interfaces.

Information system plays a very significant responsibility in e-commerce operations, e-business and enterprise collaborations and planned achievement of business. The software should be competent of supporting numerous concurrent users without compromising the performance in terms of accuracy, efficiency and speed, was made available in [69]. If no. of users is more than the performance of SPF system should not be decreased. The speed and accuracy of system should not slow down as the database grows in size and as the users are multiuser. The user should get fast response from the interface of software and the user reports should be generated speedily so that the user does not have to wait for inordinately long time to get his job done.

A design pattern for a system can be said to as set of object oriented classes within the system, related through aggregation (a-part of relationship or AND relationship), delegation, which represents an imperfect solution to a common design problem. Object oriented Design patterns are widely used to split an interface from the diverse possible implementations, to cover legacy systems i.e. technology wise very old systems or software, to include command requests, to use different platforms, and so on [70]. Reverse engineering and reengineering is required for these legacy systems. As per new requirement of technology, through the concept of re engineering we can have the legacy system operational in new environment. As we know the forward engineering is start with problem domain and the reverse engineering is start with existing software i.e. existing source code. For the purpose of software maintenance, reengineering and reverse engineering is required to sustain the software operational in new technological environment. Pattern of system calls, member functions, procedures, API, RPC and function calling can be mine from existing source code. For any web application development, these mining patterns can be used during the implementation of SPF model. After understanding the source code, desired changes can be accommodated in new version of software. As a matter of fact, the mining of design pattern information from design and source code can really assist in understanding of the adopted solution for a system, available in the literature [71]-[75]. In this work authors suggest an approach to recover design patterns from Object Oriented source code. For any web application, class diagram can be designed, and the user can generate the source code in specific object oriented programming language. Through this generated source code, developer can easily identify the pattern of operations for further implementation, was proposed by W. Ahrendt et al. in 2002 [76].

Several workers have detailed the problem of abstracting and recognizing design patterns from source code and design documents. This mined information is very useful for the re engineering of any system. During reverse engineering , information can be identify from source code, through class diagram of a system, through use case diagram and use case description, through interaction diagram through activity diagram and many more[77]. Due to IT revaluation, requirements of the use is not stable. One of the most significant issues in market is complex software development and software maintenance of complex software. As a effect of this speed there is a need to clearly know the relationships among all the diverse parts of a huge software system. Relationship between module means we have to identify the degree if interdependence i.e. coupling among interconnected modules, it was suggested by M. Gogolla et al. in 2002 [78]. Coupling should be as minimum as possible. So during reverse engineering, understanding of source code and module coupling is needed.

In object oriented programming, one of the most difficult tasks is grasping the behavior from communicating objects. In object oriented approach, system is divided into various objects and these objects communicate to each other through message passing approach, given by L. Favre et al. in 2000 [79]. The Unified Modeling Language's sequence or interaction diagrams aim at telling the message passing that takes place between various objects involved in a specified problem. Every use case has its own sequence diagram. These interaction diagrams provide a graphical demonstration of the each control transfers among different objects, was introduced in 1999 by Tamar Richner et al. [80]. For interaction diagram modeling in UML, there should be at least two objects for communication purpose. Self looping i.e. function is calling itself; we should avoid the self looping. These diagrams are helpful to represent the expected performance of a system, throughout forward engineering, but their convenience during software maintenance process is potentially even far more significant. In fact, during maintenance process the action of program understanding is necessary, and for

OO systems it may be extremely more difficult to understand the corresponding behavior of the objects [81] - [83].

Software metrics are the crucial means which assist to shrink the search space measurement: they have to be useful to the design pattern recovery process. Hence, the number of relations might be there in software. These relationships may be like aggregations, generalization, inheritances and associations etc. for reengineering we have to identify these relationships among various classes, it was suggested in references [84][85]. The metrics computed for each component class are the following:

- Identify the visibility level in the class like private, public and protected data members;
- Private, Public and protected visibility level for member functions and operations within the specific application;
- Relationship among classes like aggregation, association and inheritance (or generalization);
- Total number of data members, total number of member methods, number of super classes, and number of sub classes etc.

2.4 SUMMARY OF RELATED WORK

Three diverse work areas have been covered in this chapter like SPF. First, work on security for operating system which focuses in making operating system more flexible and easy to manage than in the past. Flexibility is added to Trusted Operating System by SPF Framework. Not only security policy [86], SPF encompasses methods to make security flexible. The SPF structure adds flexibility operations to Trusted Operating Systems. SPF involves methods to make security scheme flexible for real world computer systems and applications, was proposed in [87]. This flexible scheme of SPF will vary from one web application to another. This flexibility can be implemented on the basis of pattern mining during requirement elicitation or during requirement discovery. At the end of the requirement elicitation, all functional and non functional requirements will be clear. On the basis of this discovery process, implementation of secure software and application will become easier. Selection of necessary and un-necessary operations is the one of the most important process for the development of SPF. These necessary and un-necessary operations or system call will be related to the real requirement of user for the development of any application [88]-[91].

3

Object-Oriented Designing of Security Performance Flexibility Model

In this chapter, designing of SPF framework will be discussed. As mentioned in Section 1.4, SPF can be implemented in diverse levels of operating system. It can be implemented within individual process (Process-SPF) and system wide (System-process), it was suggested in [92]. The selection of system based SPF and Process based SPF will vary from web application to another. Operations in trusted operating system implement supplementary security will be explained in this chapter Section 3.1. SPF-System design will be covered in Section 3.2. SPF-Process design will be covered in Section 3.3. SPF-Object design will be covered in Section 3.4. This work demonstrates the implementation of SPF model for web applications in form of system and process based SPF model.

3.1 TRUSTED OPERATING SYSTEM SECURE OPERATIONS

Trusted operating system performs additional security with four major operating system subsystems. Processes, files, networking, and IPC constitute these subsystems. Table 3.1 list security relevant operation of subsystems. Security related directory operating systems are planned in separate table since they are different from normal files, was made available in [93]-[95].

3.2 FRAMEWORK OVERVIEW FOR SECURITY PERFORMANCE FLEXIBILITY

3.2.1 Operations Supported by SPF

Security Performance Flexibility may not benefit all operation mentioned in Section 3.1. Workloads and application that repeat these operations frequently are benefited with SPF. By ignoring the security checks for these repeated operations, we can increase the performance. Security checks will decrease the workloads if these security operations are executed rarely, it was suggested in, can be found in literature [8] [96]. Therefore, performance will not increase by skipping security checks on these operations.SPF framework supports the file operations, directory operations and network operations etc. Following are the operations that are supported in SPF framework.

Table 3.1: Operations supported by SPF [8]

Directory Operations	File Operations	Network Operations	IPC Operations
Change Directory	Open File	Setup Connection	Read/Receive from IPC Object
List Directory Contents	Read File	Configure Network Connection	Write/Send to IPC Object
	Write File	Send Message	
	Copy File	Receive Message	
	Seek File		
	Get File Attributes		

In 2012, Giuseppe Ghiani et al. explained that computers which are shared by users, listing directories, changing directories, copying files may occur many a times in a day. However, these operations will never be malicious in some system. Better performance can be expected for system overall by ignoring security checks for these operations, detail explanation of the same can also be found in literature [86]. Most files are publicly readable in web servers as mentioned earlier. Therefore, any security checks on them are waste of CPU cycles [87].It would be advantageous if SPF supports major file operations. Its equivalent to five additional tests on web pages. This will decrease the performance of web servers [97]. It will be advantageous if network operation in Table 3.1 will be supported by SPF.

SPF framework is also applicable to Database applications. Database applications that read random blocks on database can be benefited by adding SPF support to seek. Search latency can be decreased by skipping security checks during disk seeks [98][99]. Only operation which does not have workload that can benefit from SPF is write operation on files. Since it's added to Table 3.1 it may be useful in some systems. Computers which acts like routers benefits from SPF because storing temporary file might only be used by file system. In process operations performance cannot be increased significantly by SPF by ignoring security checks. For example, when program is executed only one security check is done [100]-[103]. It should be noted that operation in Table 3.1 are operation that advantage in the structure of SPF. Although, they may not be relevant in all SPF designs.

3.2.2 Design Assumptions for Security Performance Flexibility model

For designing of Security Performance Flexibility framework following assumptions need to be taken:-

- System call interface should be there in operating system that application uses to interface with operating system.
- Network, IPC, and Files objects are only subjects in operating system.
- For design to be correct in this chapter these assumptions are not required. They are used as reference point to demonstrate the architecture in diagram and to describe SPF Framework [104].

3.2.3 Design Overview for Security Performance Flexibility Architecture

Fig. 3.1 shows the implementation of architectural analysis of SPF model. As discussed in chapter 1, skip security checks of some system calls in goal of SPF.

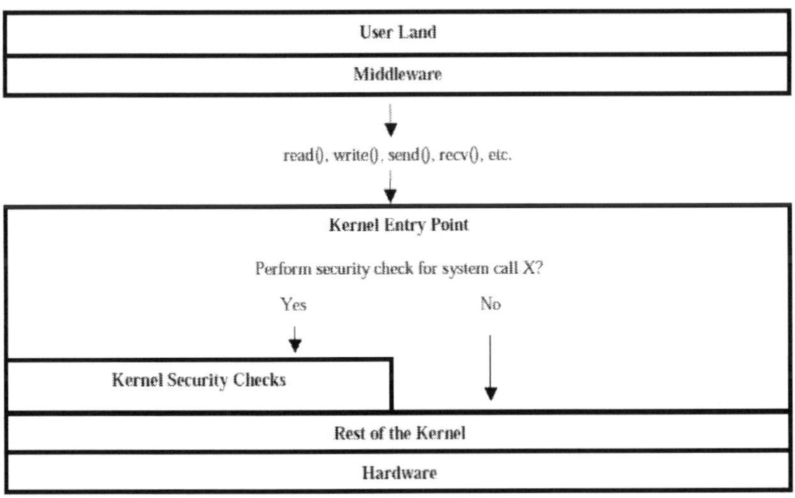

Fig. 3.1: System security performance flexibility model [133]

Fig. 3.1 raises number of question-:

- In system call X, how does the kernel knows to ignore the security check
- In System call X, how administrator does knows to ignore the security check.
- Where is information on SPF arrangement be stored

As discussed in chapter 1, SPF can be implemented in various methods. Answer to these question lies on which level SPF Framework is implemented. Design and architecture of SPF at those level and answer to question asked above would be described in following sections.

3.3 IMPLEMENTATION OF SPF FRAMEWORK AT SYSTEM LEVEL (SYSTEM-SPF)

As discussed in chapter 1, in set of system call security checks are disabled by system administrator which is allowed by System-SPF. For every application and process system calls are skipped system wide. It may benefit dedicated servers like video on demand, database, and web servers. In dedicated servers System based SPF may be useful because they have particular workloads which are executed throughout system, which can also be found in literature [8][105].

3.3.1 System Calls and Operations Supported by System-SPF

First and foremost, we should identify and determine the operations in TOS, which are supported in System based SPF. Table 3.2 shows the operations, which are supported to SPF framework. Point to be noted that table 3.2 is almost similar as table 3.1. Reason for this is that System based SPF is comprehensive to entire system and kernel. Operations that are supported in SPF model must be supported by System based SPF as well [8][106]. The necessary operations and system calls can be mined and identified at the time of requirement elicitation.

Table 3.2: Operations supported by System-SPF [8]

Directory Operations	File Operations	Network Operations	IPC Operations
Change Directory	Open File	Setup Connection	Read/Receive from IPC Object
List Directory Contents	Read File	Configure Network Connection	Write/Send to IPC Object
	Write File	Send Message	
	Copy File	Receive Message	
	Seek File		
	Get File Attributes		

3.3.2 Architectural Design of System-SPF

System-SPF structural design is demonstrated in Fig.3.2. Its configurations tell which system call should skip security checks. When call is made to operating system by application by system call interface, System-SPF Configuration should be consulted by kernel, which can also be found in literature. It tells if System call X should skip security check or not. If it indicates that security should be checked, then in kernel security layer system call should keep on to execute by doing suitable and correct security checks. After checks, assuming permission is granted, in rest of kernel system call can continue to execute, it was suggested in [107]-[111] If configuration tells that security must be skipped then kernel security layer must be bypassed by system call and continue its execution n rest of kernel. In centralized location information on System-SPF Configuration is stored, was made available in. Some administrative commands or tool have to exist so that it can adjust the System based SPF design, it was introduced by Shadab et al. in 2014 [112].

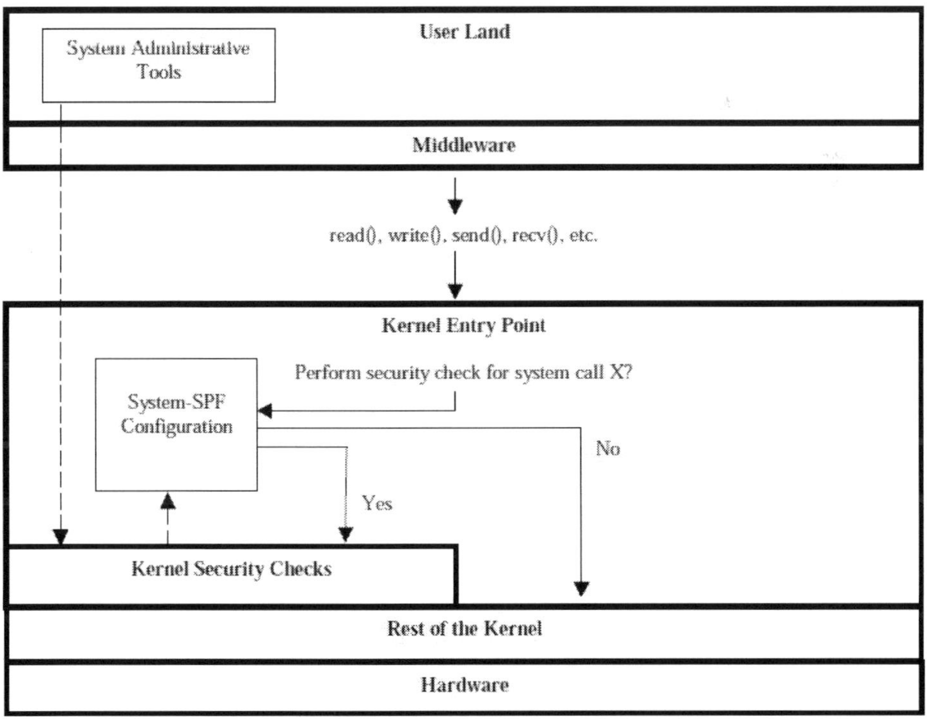

Fig. 3.2: System-SPF configuration model [134]

In Fig. 3.2, a dashed line shows the path the system administrative tools should take to change the System based SPF design. As seen to change System-SPF Configuration administrative tools have to go through kernel security layer [8]. Malicious users are not needed to change System-SPF design. So, security checks must be done appropriately.

3.4 DESIGNING OF SPF FRAMEWORK AT PROCESS LEVEL (PROCESS-SPF)

At level of process, Process-SPF permits computer administrator to manage SPF. Specific application must be selected by system administrator that should have security checks disabled in set of system call. Consider an example that host a video on demand and database server. System admin can tell that video on demand server is allowed to skip only security checks on reading files, while database server is allowed to skip only security checks on seeking files, available in the literature [113][114]. Which operation is need to be skipped and which operation is not need to be skipped is very much subjective in nature, this will depend upon the user's requirement and the particular application's requirement. The selection of these operations for a particular application is the part of software requirement elicitation process. In which we collect the user's and application's requirement, and we also represent through UML models like use case diagram, class diagram, analysis model and activity diagram etc. this modeling process is used for the development of any software or any application[133].

3.4.1 System Calls and Operations Supported by Process-SPF

Table 3.3: Operations supported by Process-SPF

File Operations	Network Operations	IPC Operations
Open File	Setup Connection	Read/Receive from IPC Object
Read File	Configure Network Connection	Write/Send to IPC Object
Write File	Send Message	
Seek File	Receive Message	
Get File Attributes		

Table 3.3 list operations or some system calls that we feel are supported in Process based SPF. Note that operation in table 3.3 and 3.1 are same. The only operation changed are copying files, listing directories contents, and changing directories. Performance

improvements are offered for these operations by Process-SPF because they are not typical of application and processes. Coping of files and directory operations are typical of general system activity and system users [134]. For example, users can read files, make copies of files, log into shared servers and move between directories.

3.4.2 Architectural Design of Process based SPF

Fig. 3.3 shows the implementation of Process-SPF. We can observe that There are various components shown in this Process-SPF figure which were no there in earlier figures. In user working area or in user interface, we have processes labelled A to C that are executing in system. The process may be related as child and parent or executing independently.

In 2013, Georgios Evangelidis et al. explained that the Design of Process based SPF is autonomous of association among various system and user processes. When it comes to hardware, we have executable program and a disk labeled M to L. Programs that initiated processes A, C, or B may or may not be executable that that on disk. Design of Process-SPF is autonomous of this information. Which system calls security check is indicated by Process-SPF Configuration. As indicated by diagram, Process-SPF Configuration is not located in the main spot, like System-SPF. Rather Process-SPF Configuration is hold up in 2 distinct areas. Firstly, each of the process running in system is linked with Process-SPF Configuration. Process-SPF Configuration A to C diagram indicates this. Second, each executable file on disk is associated with Process-SPF Configuration. Process-SPF Configuration M to L diagram indicates this. As seen in figure, Process-SPF Configuration must be precise Process-SPF Configuration as in for executable process [8], [115]-[118].

The reason for storing Process-SPF in two locations is following. Operating system compiles actual work into primary memory and respective process is created for that very program this happen when program is executed. System does not have information which signifies what exactly Process based SPF design should be available for this process. It may be saved onto disk and loaded to the system when that executable is run, was reported by S. M. K Quadri et al. in 2012 [138].

Fig. 3.3 shows that how values of Process-SPF Configuration can be spread in our system. Process based SPF design should be copied from disk to main memory after the process has been loaded to main memory. Application software summons to operating system via system call medium in System-SPF. System kernel refers Process based SPF design for process Y after application has made call to operating system. Process-SPF Configuration tells weather the respective process is permitted to omit safety checks for system call X. Therefore, if permission is granted then it can execute in remaining of kernel, if not it executes in the kernel first layer only, was proposed in [8], [119]-[122].

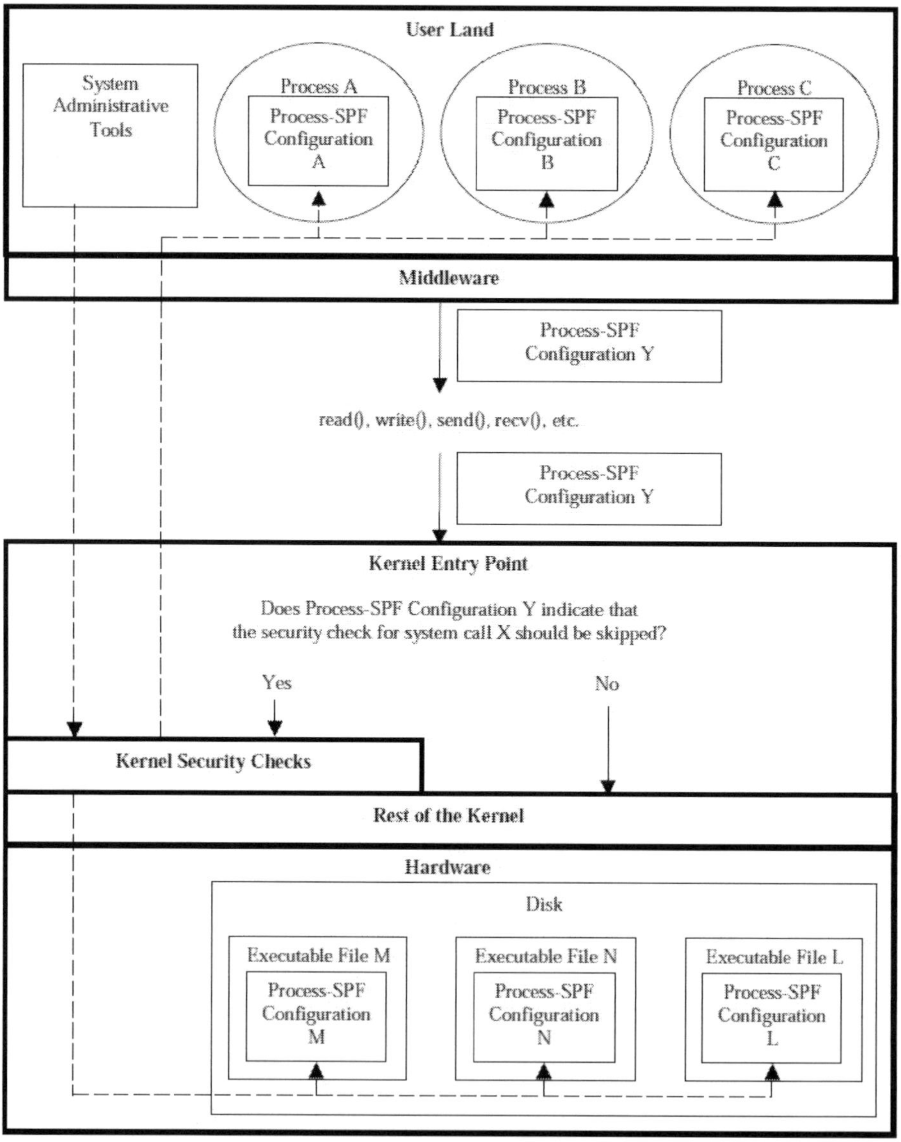

Fig. 3.3: Process-SPF architecture.

It is authoritative that kernel refers Process based SPF design or Configuration of the actual calling process which may be assumed as Y. It had been cited earlier, every process must strictly have its respective Process-SPF Configuration. Consequently, own Process-SPF Configuration should refer so as to enable system to stand on correct choice for avoiding the safety checks of kernel. Fig. 3.3 propose that Process based SPF design is send to kernel layer form application layer. However, all of this stuff is not at all needed. Requirement is such that kernel can access process Y Process-SPF Configuration, it was suggested in [123].

Fig. 3.3 shows that two important task must be performed by system administrative tools. It should facilitate system administrator power to alter Process-SPF Configuration allied with currently running processes and executable files. Given the ability of system administrators to alter Process-SPF Configuration, on presently running process it permits SPF Configuration to be dynamically updated. Within these tools suitable security should be placed, leading to no unsanctioned person changes Process-SPF Configuration in system. In Fig. 3.3 dashed lines represents the path system administrative tools must follow so as to alter the Process-SPF Configuration. Appropriate kernel security checks must be done on system administrative tools before any changes are allowed to be done to Process-SPF Configuration [150]-[151]. The marked phrases in Fig. 3.3 symbolize the road that the system administrative tools have to go through for modifying every Process based SPF pattern. The computer administrative tools must go via any proper kernel safety validations prior to any changes made to Process-SPF Configurations are welcomed.

3.4.3 Advantages and Disadvantages of Process-SPF Design

In 2017, Nitish Pathak et al. suggested that the Process based SPF has significant edge over the layout of System based SPF. Process based SPF offers administrators with great control as compare to System-SPF. In case of Process-SPF framework, the system administrator has command to choose which safety checks may be avoided for some applications and other related processes. The drawback of such strict power to control is growing administration difficulty for Process based SPF. As in the case of Process based SPF design is never primarily located as in case of System based SPF, Process based SPF needs more of hard-disk memory and more of memory to operate, was made available in [124]-[126].

3.5 BACKGROUND - CORE OBJECT-ORIENTED CONCEPTS AND OBJECT-ORIENTED DEVELOPMENT

In 1996, R. Fletcher et al. explained that Object Oriented is an ever-growing discipline and since there lies, uncertainty for what really establishes sense of object-orientation, this encapsulates the nucleus factor and defines the vocabulary to be used in the remaining of theory. Object-Oriented Development is a set of practices and approaches for manufacturing software systems grounded on real world abstractions. Object-Oriented approach delivers a relatively smooth approach for progressing from analysis, to design, through to implementation. This stays flourished by means of aiming upon the fundamental concept of the object [129]. Object oriented development is presently the most accepted software development style. UML has now become the language of choice intended for developers who wish to imagine and model the system underneath progress. UML is used in many customs for expressing the concepts such as software specification, website structure and business modeling. In the proposed design and development, we are using UML2.0 based conceptual modeling for the development of secure application [130].

Modeling helps in visualizing the functionality of software systems and simplifies its design process further. By the help of the suitable models, organizations are able to understand the structure and overall functionality of any system, by identifying flaws in the system and applying required modifications at an early stage of its development. UML is a standard language for modeling the software system. It's a graphical language and a framework for visualizing, specifying and documenting the artifacts of software and non-software systems. It makes the programming simpler, more effective and manageable, explained by Martin F. et al. in 2013 [1]. UML was introduced by three eminent scientists named James Rambaugh, Ivar Jacobson and G.Booch and was released by the Object Management Group in 1997. Until now, many versions of UML have been released, each with improvement over its previous version. Object oriented design uses objects in the programming language which helps in modeling the patterns of any system. It manages the software complexity and also enhances usability and maintainability of software; it was suggested in [127]-[128]. Because of increasing software's ability to trace and collect a large amounts of data and performing mining operations to obtain useful knowledge or information, which is helpful in determining future business policies and useful patterns. There is also a need to develop the system to prevent disclosure of sensitive information or patterns.

For this purpose, Privacy Preserving Data Mining has now become a demanded and proven methodology for most of the organizations [152].

3.5.1 Objects

An object is simple something that has SBI properties. Every object has its own states, every object has its own behavior and every object has its own identity. This is relatively dissimilar since previous procedural and structural approaches. In procedural language, program is divided into various functions but in object oriented approach, whole program is divided into various objects. These objects communicate through message passing approach. The data structure of an object is denoted by a arrangements of variables known as attributes which have an identifier (name) and a data type which can either be vital (such as integer or float) or user defined. During the analysis at analysis phase, analyst must identify various objects like control objects, interface objects and entity objects within the particular web applications [153]. Interface object in a system is used to provide interface for user, so that end user can perform their desired task very easily. Control object is used for the purpose of security checking, validation operation verification. At the last entity object of analysis model is used as data base to store the information. After identification of these three objects in specific system, analyst designs the analysis model as a part of forward engineering.

3.5.2 Classification

A class is a collection of similar object instances. Class follow the concept of date abstraction, which means only relevant information will be displayed to user. Class fulfills the concept of encapsulation. Encapsulation means binding the data members with its associated member function within the single unit that is class. In this book we will design the class diagram for some real life case studies, as discussed by Ahn Y. et al. in 2003 [127].

3.5.3 Identity

Each and every individual object is a distinct entity that able to be recognized from all other objects in the system. In other disputes, objects can distinguished even if they are built upon the similar class and share some identical values for all of their attributes. With OO databases, uniqueness are often, more obvious with the presence of inimitable physical object identifiers, it was suggested in [138].

3.5.4 Inheritance

Inheritance supports the concept of reusability. In modeling term this is also known as generalization. Where a base class is inherited by derived class. These base classes are also acknowledged as a parent class and super class. And derived classes are known as subclasses and child classes.

Each subclass receives the attributes and methods of its super class and enhances its own specialized performance. The ability to assign commonality in a hierarchical means can decline the quantity of repetition in a project. In exercise, popular OO languages such as C++ and Java afford an amalgamation of special practices like single, multiple, implementation or interface. In modeling term, inheritance is known as generalization. This means a kind of relationship or is a relationship among the object oriented classes.

3.5.5 Polymorphism

In its over-all sense, polymorphism describes just that method with having the same name that may act differently in different classes. The purpose of polymorphism is same thing for performing different task. This can be run time and compile time polymorphism. This type of polymorphism usually associated through object-orientation is dynamic polymorphism, that includes late or runtime binding. Function overloading and operator overloading is the part of compile time polymorphism or early binding, is explained in [129][130].

3.5.6 Encapsulation and Information Hiding

The terms encapsulation and information hiding are repeatedly used synonymously though to be precise, data hiding is used as a technique for attaining encapsulation. Encapsulation means binding the data with its associated member functions within the single unit. Information hiding is applied in OO languages as the aptitude for a class to specify which attributes and methods are hidden from additional classes. For example, in Java and the C++, this is completed by public, private, and protected keywords [131][132].

3.5.7 Benefits of an Object-Oriented Approach

Hao-hua Chu et al. in 2015 explained the specific meaning of abstraction: "An abstraction signifies the essential features of an object that discriminate it from all other types of objects and hence provides crisply defined theoretical boundaries, qualified to the perspective of the watcher." This explanation clues at one of the welfares that OO development provides relation to abstraction [4].All object oriented features are very much useful in web applications for forward engineering, reverse engineering, reengineering and roundtrip engineering, is explained in [133].Reusability, Data Abstraction and Encapsulating classes etc. are some advantages of an Object-Oriented Approach.

3.6 DESIGN PATTERNS

Though design patterns are not severely talking point in an OO concept, they are enclosed here because of their growing occurrence and approval within the OO community [93]. Designs capture actual design experience and methods that have been proven over a number of projects, pattern knowledge is often haul out using a method known as pattern mining, is explained in [95]. The mining of operations can be identified during the requirement elicitation phase of software development life cycle. After identifying the system call patterns through requirement engineering, implementation of SPF model for web application is possible.

The pattern of system call and operations in a particular application can be identified from the existing source code of open source software. This source code can be modified according to our latest requirement; it was suggested in [134]. Through the source code in specific programming language, Complexity measures can be identify like information flow between module, McCabe complexity, lines of code, Halstead complexity, data matrices, etc. have been widely explored [57, 58]. The performance of this system will be dependent on the attributes which we are selecting during the development from mining in various web apps. After discovering these system calls or operations through requirement engineering or requirement elicitation, at analysis phase, we can implement these system calls on OS level so that the security and performance can be maintained equally for all the web apps, it was suggested by Nitish Pathak et al. in 2017 [135].

The selection and observation of necessary member functions, operations, system calls and data members will vary from one application to another. Various system calls, operations, patterns and rules needs to be discovered by the web developer that can be used to improve the system performance of software.

3.6.1 Object-Oriented Analysis

OOA is primarily worried with the problematic domain i.e. what the system must do, not how it would be consummate. The following is a list of approximately the activities that happen during OOA:

1. Recognize limitations and responsibility for the system.
2. Launch the user-requirements and needs.
3. Discover scenarios then use-cases of the planned system.
4. Recognize classes from problematic domain entities, and describe their their associations with each other.
5. Inspect the related systems and trace existing software mechanisms for reuse, is explained in [102].

Object-Oriented Design

OOD is concerned about the appropriate and effective structuring of a complex system by means of an OO notation at a higher level than the individual constructs of a programming language. Some of the activities performed during the OOD phase are:

- Explanation specific classes, which have no equivalent in the problem domain, are further added to the design.
- Modules are well-defined with specific devotion to cohesion and coupling.

Object-Oriented Programming and Multi-disciplinary Context of the Book

During the OOP phase, the classes and relationships labelled during design are interpreted into an executable form through a precise OO language and/or database system. For code generation and forward engineering, we can use any object oriented programming language like C++, Java and VC++ etc. here in this work, we are using C++ for round trip engineering.

Here, we are worried about the detention of theoretical entities (or classes), the relationships prevailing among them, and the logical assertions that restrict, or constrain, the contribution of unspecified entity or class instances in relationships held among it and adjacent ones, as signified in one or more theoretical model diagrams[133]. Now a time there is a growing dependence on web applications, reaching from characters to big organizations. Approximately the whole thing is stored, accessible or traded on the web. The omnipresence of net applications in our means of life and in our economy is so important that it brands them a usual target for mischievous minds that want to adventure this new streak [136]. The security inspiration of web application developers and commissioners should reflect the importance and significance of the properties they are supposed to defend. Nowadays UML deliberate as a de facto standard in software development and is used in numerous areas ranging from scientific modeling to business modeling [137].

Web applications, which are first conceived to let the user-to-system dialogue, are lengthy with Web services, which allow system-to-system interaction, and with procedure control primitives, which permit the implementation of the required business constraints. Models are industrialized using different kinds of languages and/or notations. We have surveyed approaches in the literature speaking dependability modeling and the examination of software systems stated with UML. Development of web applications that combine existing web facilities becomes more and more general. In this book, we have presented the application of such user-centric applications of model-driven methods for the development [138].

The reverse engineering is performed to sustain legacy code. That's why; it should be focused on system maintenance and not on program understanding. This should be done in such a way that we won't have to do the reverse engineering of the system again and again after every modification made to its code over time. So, we have to focus reverse engineering on design recovery. The recovered design need to be based on formal methods as a sophisticated tool is needed by us to perform system maintenance efficiently. It has been seen by us that to recover such a design, transformational reverse engineering is a possible approach. It means that the same domain knowledge and infrastructure of transformational forward engineering is needed by us that, during the systems' construction, can provide us with a design [139].

Model Driven Development (MDD), that makes use of models to specify programs, is a rising standard for the development of software. It also uses model transformations to synthesize the executables. By composition of features for creation of models, and then by the transformation of these models into executables, the MDD shows how products in a software product line can be synthesized. The WWW used to be a small work community two decades ago and now it has grown to become the major and mainly accepted infrastructure for search, communication, social interaction, information distribution and commerce. Handling different devices, different programming languages (Java, Objective C, and visual C++) and multiple operating systems (Android, Apple iOS, windows mobile) are few challenges that are involved in mobile application development. New software engineering tools and techniques start gearing towards the mobile platform as number of mobile applications increased and smart phones became a part of life[140][141].

It was observed in this book that the Object-Oriented decomposition and the Software modularization in particular is and approach for the understanding of the source code, for the improvement of the organization. A well-connected demonstration of the classes, which make the software system, is needed to the created by the software engineers in order to understand the Object Oriented software. The relationship among the classes also needs to be understood individually along with the each class individually [117]. Creation of a system where classes having low coupling and high cohesion is one of the purposes of Object Oriented analysis and design. The properties of these classes facilitate testing, maintainability, comprehension, reusability etc. Relationships between objects and a common structure are specified by the class diagrams. These class diagrams are often detailed by the constraints. In a series of steps, the occurrence of System Specification, Design and implementation takes place [154].

A software product's long term success is dependent on the high quality of system architecture. Implementation and design of an efficient, usable, secure, reliable and maintainable software system is needed to be done by the software engineers. These software engineers need to design the architectures that can judge the quality tradeoffs, fulfill the quality requirements and ensure that the implementation follows the architecture. They also have to evaluate, monitor and improve the quality continuously throughout the SDLC in case of long living software systems. A key role is played by Models to support software engineers in mastering these tasks [142].

A widely accepted OO system design and modeling language called UML has recently been used for the description of web system recently. This language has been used for several years for descriptions of software architecture. UML extension has been used by several work groups to describe the architecture of web system. In book e-commerce applications design, a process of software production has been described. An OO web-solutions modeling approach (OOWS) drives this process. Methods for web-based application development are provided in this approach. In the proposed process, we completely specify a system using OO conceptual modeling techniques to properly capture the definite functionality of an application related to e-commerce [143]. The implementation of web solutions is required by more and more organizations with SPF based high security and functionality to execute business transactions on the Web.

4

Implementation of Object Oriented Secure Round Trip Engineering for Web Applications

An important problem in today's Internet is Web application security. Many programmers do not have adequate knowledge about secure coding is the major cause of this status, so they leave applications with vulnerabilities. This work proposes SPF based secure software analysis, SPF based secure software design, SPF based secure software development and roundtrip engineering based maintenance approach.

We have also demonstrated how real-life product-line products can be built on top of this platform. This work creates an experiential base to showcase the effectiveness and efficiency of SPF based object oriented modeling mechanism in software development, with primary focus on security in applications. To retain the balance among security and web application performance, we propose amalgamation or combination of security based forward engineering and round-trip engineering. This round-trip engineering in software development is used to maintain synchronization between source code and software design.

Various pattern mining approaches have been proposed and they are used for different applications such as moving object tracking, network monitoring, scientific data processing, financial or medical data analysis, etc. Discovered patterns were useful in different contexts to predict data behavior (or trend), to detect anomalies, or in general sense, to improve system performance or to simplify data processing. However, in the context of web archiving, patterns have never been used to the best of our knowledge. Web Archiving is the process of continuously collecting the portions of WWW and preserving them for the future generations. In this book, it has been shown how patterns of operation and system calls can be identified for the development of applications. It is due to the distribution pattern that has a direct impact upon the QoS of the composition. However, a considerable amount of code is required to realize the patterns of these distributions. We should identify the system call and operation patterns in a particular application. Here we are considering some case studies for the same.

4.1 OBJECT ORIENTED SECURE ROUND TRIP ENGINEERING FOR WEB APPLICATIONS-1

During the development of web-based applications MDD have become more important due to the functionality of working together of a variety of user types. In this book, to support a feasible way of helping developers in taking into account issues related to this variety of users in the Web, author describes the adoption of MDD approach. Generally, time consuming tasks and high cost is implied by a good design of interaction and communication issues, and requires developers to rapidly change the conception of the interfaces and to be flexible. Communication requirements and interaction mapping are covered in the meta-models approach that has been presented in this book in a useful and a practical way during the web application development.

Integration of presented web services has become more and more accepted in the development of new web applications. This type of web applications can be classified as the composition of web services around UI flow. In this book, for the development of user-centric web applications, the author has presented the application of model-driven techniques. Modeling of application using UML class, transformation to appropriate code (platform specific) and state transition diagrams has been presented in this approach. Since the rise in the pervasive access to internet, and in the usage of mobile code and networked appliances, there has been a sudden increase in the importance of the protection of software. In today's internet world, it has become significantly difficult to make a web application safe and secure due to a rise in the ability of software to collect and trace huge amounts of data and to perform mining operations in order to obtain useful information or knowledge. This is helpful in determining the useful patterns and future business policies [155]-[156].

Various programmers don't have enough understanding of secure coding programming languages. This is the main reason of this condition. So, major applications are left with a lot of bugs and vulnerabilities by them. In order to resolve this difficulty, we can use source code analysis to discover these vulnerabilities and bugs. In this work book, we have investigated the use of a fusion of such methods to identify the bugs and vulnerabilities with fewer mistakes. We know that the trusted OS are better for the implementation of web apps for maintaining more security. A centralized security methodology has been implemented by the secure web applications project across the entire web project.

In today's internet world, it has become significantly difficult to make a web application safe and secure due to a rise in the ability of software to collect and trace huge amounts of data and to perform mining operations in order to obtain useful information or knowledge. This is helpful in determining the useful patterns and future business policies. Various programmers don't have enough understanding of secure coding programming languages. This is the main reason of this condition. So, major applications are left with a lot of bugs and vulnerabilities by them. In order to maintain more security, for the implementation of software, trusted systems are a better choice. To maintain high security in web projects, these trusted OS may prove to be more preferable choice.

In Fig. 4.1.1, we have verified the blueprint of Trusted OS for web application. The better quantity of security is placed into the kernel of the operating system in case of trusted operating systems for web apps. In Fig. 4.1.1(b), we have showed this safety and security with a significantly thicker layer of security checks in kernel. As shown in the Fig. 4.1.1(b), in trusted operating systems, the kernel safety checks are much more than in standard operating systems. These security checks make these operating systems much more secure. But due to this thick layer of safety and security, the trusted operating systems bear performance issue in small web applications. Before any valuable or important work is performed by any system call, it has to go through this big layer of security checks. These web applications may bear from performance issues due to these security checks, if we use trusted operating systems.

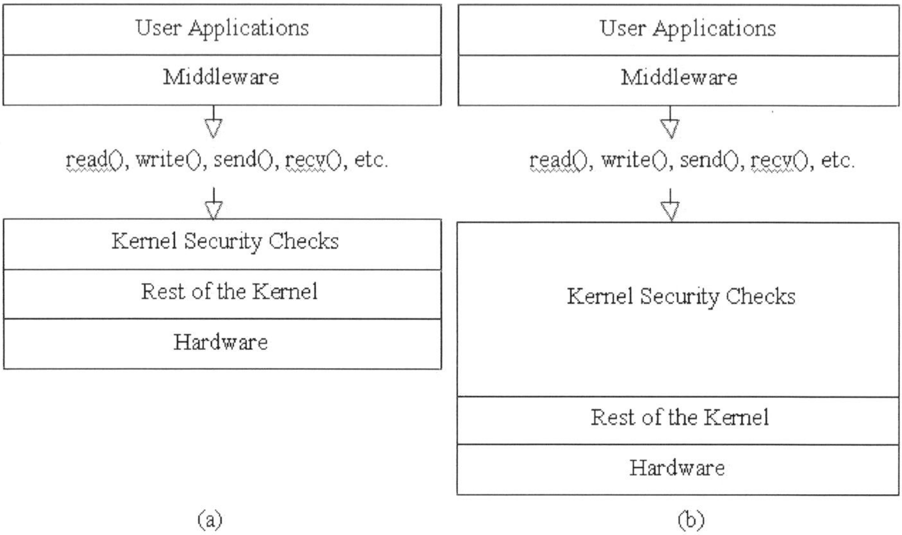

(a) Ordinary Operating Systems (b) Structure of proposed trusted Operating Systems for web applications

Fig. 4.1.1: Structure of trusted operating systems and ordinary operating systems for case study 1

The performance of this system will be dependent on the attributes which we are selecting during the development from mining in various web apps. After discovering these system calls or operations through requirement engineering or requirement elicitation, at analysis phase, we can implement these system calls on OS level so that the security and performance can be maintained equally for all the web apps.

There has been seen a tremendous growth in software industry in last few years. In this current growing information technology trends, organizational endurance depends on customer's satisfaction and the quality of the software product delivered to them. Apart from this success factors, many of the software projects fails due to being schedule overrun, cost overrun or due to lack of conformity to quality standards.

In 2006, Johnson Rumbaugh et al. explained that a good quality software project can be produced only if the requirements from the client is gathered and understood effectively. Understanding of software requirement of the client is very crucial stage of every project development. Various workers have been worked with the objective to find out effective requirement engineering processes to produce quality products on time within a specified budget. Literature reveals that requirement engineering process depends on the technique employed to carry out the elicitation of the requirements. It is necessary to use application specific elicitation technique for requirements elicitation. Requirements elicitation techniques collect requirements from the users or stakeholders and communicating those requirements to system specialist. The successful implementation of any software system is highly depending on the quality of requirements and the elicitation technique used to gather that requirements. The objective of this study is to find out user requirement for a particular application through requirements [10].

The study also emphasis on taking the evidence from the IT professionals for finding out the effectiveness of the elicitation techniques in terms of cost, popularity, time etc. Many workers are working in the area of requirements engineering. Standish Chaos Report 2014 shows, 61% software projects are affected from cost overrun, time overrun and improper services due to underprovided user involvement, incomplete requirements, dynamically varying requirements specifications, ineffective requirements and software product mismanagement. Workers says that security should be provided 'since the inception' of software system. This study focuses on the first phase of requirements engineering and make use of questionnaire tool to get the different viewpoints of requirements engineering practitioner for identifying the application specific elicitation technique in industrial level.

The object oriented model driven approach has been explained in this book along with structure to Web application designing and secure execution environment. One more major concern of web applications is the security of web application, that's why Security Performance Flexibility (SPF) has also been focussed in this book. A developer must balance Security during all phases of SDLC as it is a major concern for web. It must be recognized by you that how significant the security is for your software project.

A trusted operating system will be slower than standard operating system due to additional security checks in the kernel that has been demonstrated by the Fig. 4.1.1(b). If these OS are used by us for web apps, then we get lower efficiency and lower performance, but more security features for web apps, what's difficult is the balancing of the performance with the security parameters of this OS for desired web applications.

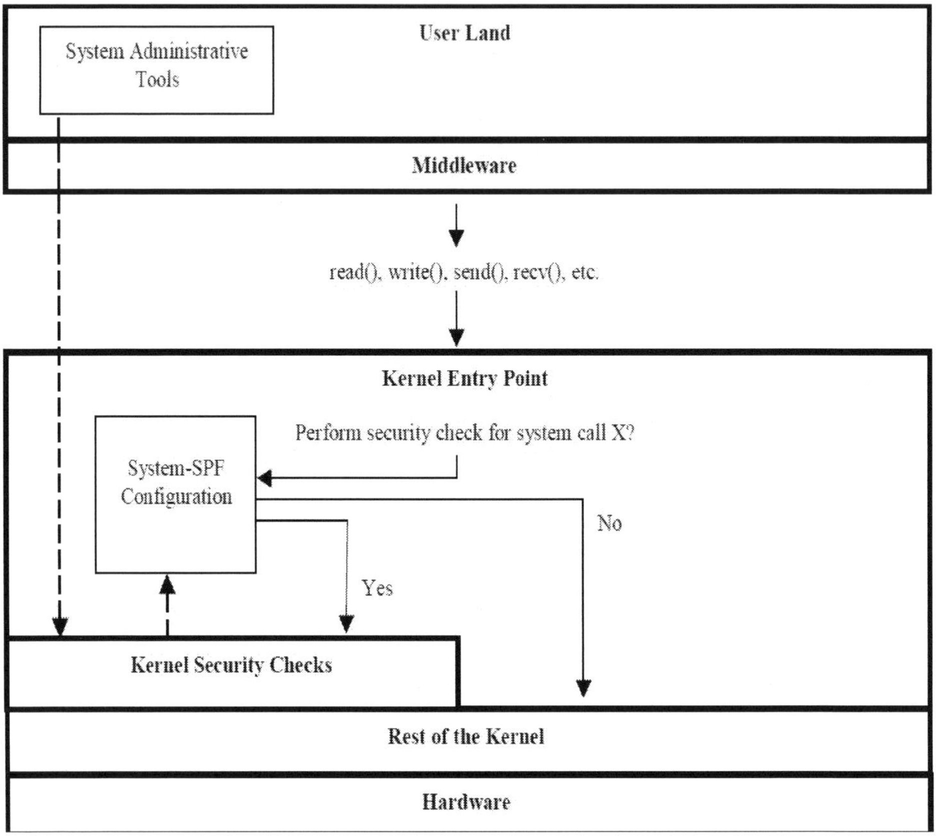

Fig. 4.1.2: System-SPF structural design for stock control web application

Only a little amount of parts of OS security is actually necessary, as far as computing workload is concern. The selection and observation of necessary member functions, operations, system calls and data members will vary from one application to another. Various system calls, operations, patterns and rules needs to be discovered by the web developer that can be used to improve the system performance of software. Some unnecessary parts of security for web applications can be skipped by the administrators as per the SPF, according the Fig. 4.1.2. In order to balance the performance and security needs in particular web application, developer and designer can use the SPF structure during the development of secure operating system. Performance, efficiency and quality of web application can be increased in all respect whenever some unnecessary and useless parts of the trusted operating system security are prevented. To achieve such development and designing, first of all, the part that can be disabled, in order to get the highest performance in a particular web application, is recognized. Architectural construction behind SPF configuration has been demonstrated in Fig. 4.1.2. The blueprint for a secure web server can be developed after OO modelling. After this, the implementation and run can be done on the system SPF based trusted OS. The administrator of a web server can use SPF model of OS to turn off all safety checks and read security checks in a web server.

Table 4.1.1; table 4.1.2 and table 4.1.3 show the performance results that are appropriate to SPF. These tables show the results of with SPF, without SPF and show the performance compression.

Table 4.1.1: Security checks executed in SELinux trusted operating system

File System Tests	SELinux without SPF	SELinux with System-SPF model
Random Disk Reads (K) per second	94167	93135
Random Disk Writes (K) per second	79188	79508
Sequential Disk Reads (K) per second	335527	325591
Sequential Disk Writes (K) per second	149616	153174
Disk Copies (K) per second	102252	102744

Table 4.1.2: Security checks skipped in SELinux trusted operating system

File System Tests	SELinux without SPF	SELinux with System-SPF model
Random Disk Reads (K) per second	94167	99762
Random Disk Writes (K) per second	79188	84768
Sequential Disk Reads (K) per second	335527	363571
Sequential Disk Writes (K) per second	149616	159727
Disk Copies (K) per second	102252	110315

Table 4.1.3: Comparison of performance improvement after security checks skipped in SELinux trusted operating system

File System Tests	SELinux without SPF Performance Degradation	SELinux with System-SPF model Improvement Over SELinux No SPF
Random Disk Reads (K) per second	-6%	+5%
Random Disk Writes (K) per second	-6%	+6%
Sequential Disk Reads (K) per second	-9%	+7%
Sequential Disk Writes (K) per second	-5%	+6%
Disk Copies (K) per second	-7%	+7%

The choice of System-SPF will depend upon the requirement and complexity of web application. In this article, we are using the experimental software designing and development for Store Stock Control. UML 2.0 is for the same purpose. No doubt, that software design and development process and object oriented empirical modelling is very important for the development of e-commerce applications and IT applications. We will consider the subsequent requirements for Store Stock Control system: Store Stock Control system requires a new point of sale for their several stores all through the United Kingdom to replace their old mini store systems. A sale associate or assistant will be capable to process a product order by entering a unique product numbers (This product No. will be unique in stock control system). The Store Stock Control system will show a description of product, price of the product and available stock respectively. In stock products will usually be collected directly by the consumer or customer from the Store Stock Control system but may be selected for delivery to the consumer's home address. The delivery of product at home will be chargeable; there will be a charge for the same. If required or requested stock is not available, the sales manager or sales assistant will be able to make a backorder for the required product from a nearby warehouse or from a regional warehouse. The required and requested products will then either be delivered express from the regional warehouse to the consumer's home address or to the store for collection by the consumer.

The projected approach for above case study is based on the following phases of the development process:

- First of all, extract web application requirement and requirement discovery through requirement elicitation techniques and Requirement Engineering. This process will help in mining pattern for the system calls.
- Design the use case diagram according to mining pattern of system calls, to represent the functional requirement of web application for the analysis phase of SDLC.
- Explain almost every use case in a textual approach i.e. description of major use cases is required for more refinement. If something is not clear through use case diagram then use case description will be very beneficial at other phases of software development life cycle.
- Specify process behavior using sequence diagrams.
- Designing of class diagram and implementation of Forward engineering in C++ source code using IBM Rational Rose for the same secure web application i.e. design phase Concept of reverse engineering, re engineering and round trip engineering for the same secure web application.

The Use case diagrams have been drawn to emphasize all important scenarios and cases regarding the system use by these actors. As we know, 'Use Cases' are used to represent the functional requirements and services of the software system. These functional requirements are implemented through functions, member functions, methods, procedures, sub routines etc during implementation phase of software development life cycle. So, we should design the use case diagram of above case study for round trip engineering.

In Fig. 4.1.3, we can see that store manager, sales assistant, customer, warehouse person and stock manager etc. are the actor. Rest of the diagram indicates the use cases (for example Display Stock Details, Payment Use Case, Order Delivery Use Case and Sales Summary Report Use Case etc.) i.e. functional requirement of Store Stock Control system and their relationships.

The Fig. 4.1.3 shows the generalized use case diagram for Stock Control system.

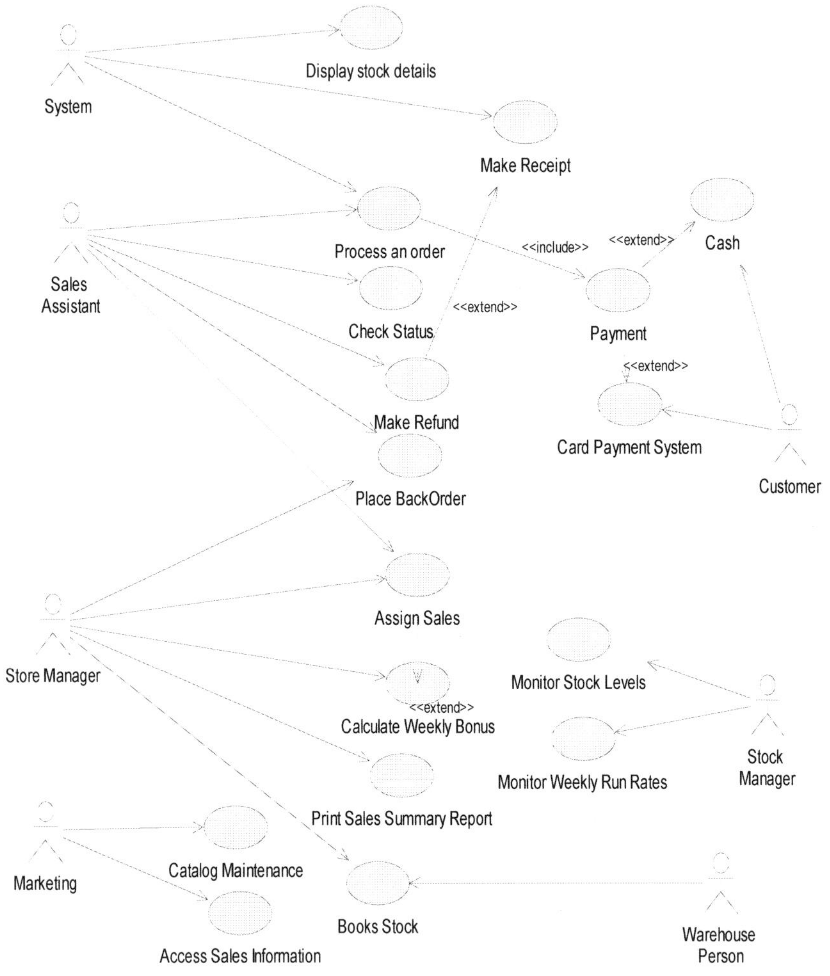

Fig. 4.1.3: Generic use case diagram for store stock control based web application

Tables 4.1.4 to 4.1.12 present the use case description of major use cases for Stock Control system.

Table 4.1.4: Use case description of display stock details

Brief Description	**The system displays info to Sales assistant**
Actors	System
Flow of Events	• Sales assistant enters the product numbers (Pro. No.) and required quantities into system. • System displays the description of the product.
Alternative Flow	If the stocks control system in not functioning, it will not start.
Precondition	Product numbers must be entered
Post condition	Information regarding product is displayed

Table 4.1.5: Description of order delivery use Case

Brief Description of delivery	**This enables the customer to get the delivery of their order from the stock control system.**
Actors	Customer
Flow of Events	• Places order • Enter delivery details
Alternative Flow	If the processing doesn't get successful, generate error report to the sales assistant.
Precondition	The order should be successfully processed on the system
Post conditions for order delivery use Case	If the use case was executed successfully, the order shall be delivered to the customer. If the order is not delivered, the system state will remain unchanged.

Table 4.1.6: Payment use case description

Brief Description of Payment	**Payment Use Case allows the customer to make payment for his order in the stock control system.**
Actors	Customer
Flow of Events	• Places order • Enter delivery details • Makes payment
Alternative Flow	If the processing doesn't get successful, generate error report to the customer and order is unsuccessful.

Brief Description of Payment	Payment Use Case allows the customer to make payment for his order in the stock control system.
Precondition	The order should be successfully processed on the system
Post condition	If the Payment Use Case was successful the payment for the order shall be made by the customer. If not, the order is incomplete.

Table 4.1.7: Credit card payment use case description

Brief Description	The credit/debit card is the mode of payment for the customer.
Actors	Customer, Sales Assistant
Flow of Events	• The customer pays for the goods with credit card. • The card payment is verified using online transaction system.
Alternative Flow	If the processing doesn't get successful, generate error report to the customer and order is unsuccessful.
Precondition	The order should be successfully processed on the system
Post condition	If the Credit Card payment use case was successful the payment shall be made by the customer. If not, the order is incomplete.

Table 4.1.8: Check status use case description

Brief Description	The sales assistant will be capable to check the status of the available stock.
Actors	Sales Assistant
Flow of Events	The sales assistant confirms the status of the available stock.
Alternative Flow	None
Precondition	None
Post condition	If stock is not available back order is placed.

Table 4.1.9: Cash payment use case

Brief Description	This use case enables the customer to make payment for his order in the stock control system through hard cash.
Actors	Customer
Flow of Events	• Places order • Enter delivery details • Makes payment through cash

Brief Description	This use case enables the customer to make payment for his order in the stock control system through hard cash.
Alternative Flow	If the processing doesn't get successful, generate error report to the customer and order is unsuccessful.
Precondition	The order should be successfully processed on the system
Post condition	If the use case was executed successfully, the payment shall be made by the customer. If not, the order is incomplete.

Table 4.1.10: Make refunds use case description

Brief Description	The sales assistant makes the refunds to the customer by initiating this use case.
Actors	Sales Assistant, Customer
Flow of Events	• The customer produces a valid receipt. • The refunds are made by the sales assistant. • The use case ends.
Alternative Flow	• Invalid Receipt • The customer doesn't produce a valid receipt. • He is asked for a valid receipt. If he is able to produce a valid receipt, the basic flow step 'REFUND' is resumed. Otherwise the use case ends.
Precondition	Customer has made all the payments.
Post condition	Customer has got refunds.

Table 4.1.11: Sales summary report use case

Brief Description of Sales summary report	This enables the store manager to view the summary report of sales
Actors	Sales manager will be actor to generate the sales summary report
Flow of Events	• Logs into the system • View summary report
Alternative Flow	If the processing doesn't get successful, generate error report to the sales manager.
Precondition	The stock information along with sales details should be present on the system
Post condition	If the use case was executed successfully, the summary report is delivered to the sales manager. If not, the error is generated.

Table 4.1.12: Monitor stock levels use case

Brief Description of Monitor stock levels	**This Monitor Stock levels Use Case enables the stock manager to monitor the stock levels**
Actors	Sales manager
Flow of Events	• Logs into the system • Monitor stock levels
Alternative Flow	If the processing doesn't get successful, generate error report to the stock manager.
Precondition	The stock information along with sales details should be present on the system
Post condition for Monitor stock levels	If this use case was successful the stock levels is delivered to the stock manager. If not, the error is generated.

For novel modeling and full UML notation, during analysis phase of software development life cycle, various predefined stereotypes can be used during the designing of use case diagram. Stereotypes show the dependency between two use cases. These stereotypes vary from one diagram to another. For example <<include>> and <<extend >> stereotypes are used in use case diagram and create and destroy stereotypes are used in sequence diagram. These stereotypes provide additional information about the system. Both the stereotypes show the dependency between two use cases, as, we can notice in Fig. 4.1.3. This solution can be complemented by designing a sequence diagram. For each and every use case, a unique sequence diagram can be designed. The interaction or sequence diagrams are used when we wish to know the message flow in the software systems.

During the design of sequence diagram, we can have synchronous and asynchronous messages between objects to show the calling sequence of member methods. In any software or in any applications Synchronous messages are used when situation is related to stop and wait. The selection of synchronous and asynchronous messages will vary from one application to another application. Currently, message flow means the sequence and ordering of flow from one object to another object. That is why sequence diagram is used to capture dynamic environment but from a special viewpoint. Fig. 4.1.4; demonstrate the sequence diagram for Store Stock Control system. The intentions of these dynamic diagrams are to imagine the interactive activities of the software system. Nowadays visualizing interaction is a difficult task. Thus the solution is to use special types of models to capture the diverse aspects of the interaction. The usages of interaction diagram are as follows:-

- To confine dynamic performance of a software system. As we know, in object oriented language, object will communicate through message passing approach.
- To express the communication flow (API, RMI, function calling, and message flow etc.) in the software system.
- To clarify the structural association among the objects.
- To illustrate communication and interaction between objects.
- No. of Objects taking part in the interaction during interface.
- Message flows (synchronization and a synchronization etc) among the objects in web application.
- The sequence or ordering in which the messages are flowing within the web application.
- Object organization of software systems.

During implementations, Web Developers can implement these Message flows, Object organization, structural association of the objects, ordering of member functions, forward engineering in application, reverse engineering and round trip engineering of software systems. The generic Sequence diagram of projected case study is shown in Fig. 4.1.4:–

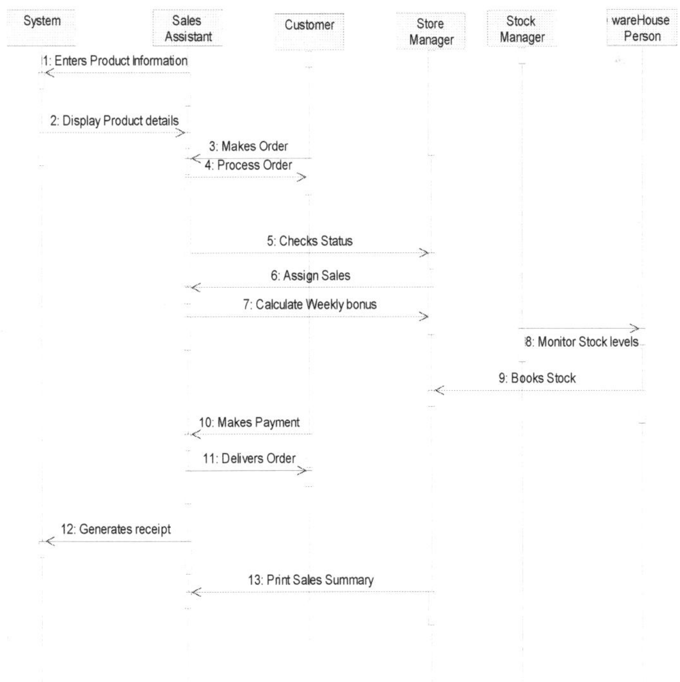

Fig. 4.1.4: Sequence diagram for store stock control based web application

This book focuses on empirical relationships between SPF model, the use case, sequence diagram and class diagrams. Web designer and developers need to make a well-connected object oriented classes, where classes have high cohesion and low coupling between them. Degree of coupling between classes should be as minimum as possible. Classes have low coupling means the dependency between classes should be as minimum as possible. With the help of above use case diagram, we develop the component based classes. In object oriented class diagram, designer will identify the classes. These classes can be identified through software requirement specification (SRS). As normal practices, actors of use case diagram are considered as classes and the use cases are considered as member functions or methods of the classes. After this we generate source code in C++ and execute forward engineering and reverse engineering for the same web application with the help of forward engineering tool. When we want to model the structure of a system or a web application, we can make use of object oriented class diagram. Classes of applications are more or less like entities in entity relationship diagram. When we want to model the interaction among objects in runtime, and sequence of function calling and method invocation, we can design and draw the interaction or sequence diagram for major use cases [27]. As we know, analysis is close to design phase of software development life cycle and design is close to development.

Class diagram obviously shows the mapping with object oriented languages like Java, C++ etc. These object oriented languages are close to real world mapping. So from realistic experience class diagram is normally used for construction purpose of software development process. So in a brief, as we know the class diagrams are used for describing the static analysis of the software system and showing the association and multiplicity between the elements of the static view. As source code is in text form, it is difficult and is hard for human to read or analyze, especially when the logic is difficult and involves a large number of classes. “A picture is worth a thousand words”, by visualizing source code with diagram, we can simply realize the classes involve as well as their association in run time. so that we can gain a better understanding of C++ source code by reading diagram instead of looking to a possibly thousand lines of source code. The standard class diagram of Store Stock Control web application is as follows in Fig. 4.1.5: –

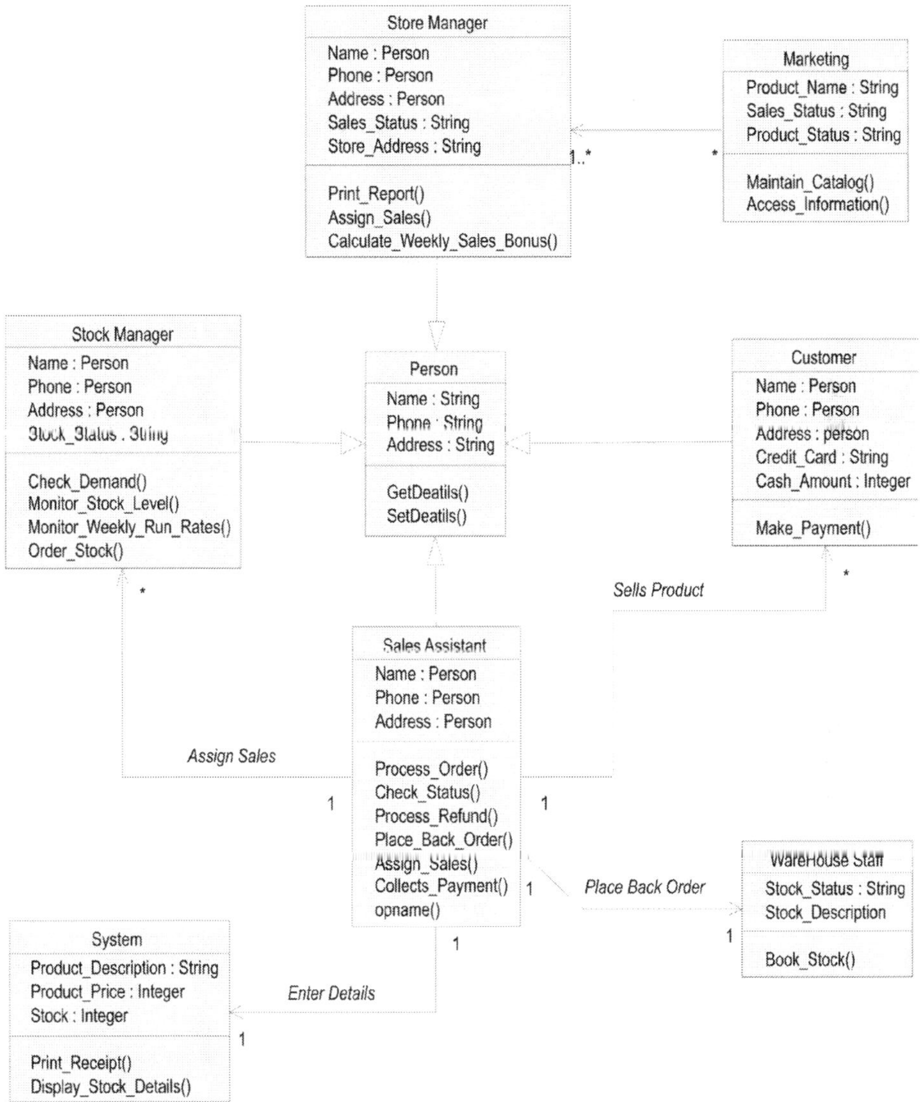

Fig. 4.1.5: Generic class diagram for online store stock control web application

Logical and technical Designers can build a high level object oriented class model, then pass to web developer to perform more lower-level software system modeling, and ultimately follow the forward engineering to generate source code. With IBM rational Rose, we can simply generate source code in object oriented language from class diagram. This source code will work as blue print for software development team.

In this Store Stock Control Based Web Application, storing objects may be sales clerk, inventory, Credit card, Cheque, store manager, Payment, Person, marketing, stock manager, person, warehouse person, invoice, system, customer etc. (See Fig. 4.1.5). The class wise equivalent C++ code of this case study is as follows-

Code:

```
#ifndef CUSTOMER_H_HEADER_INCLUDED_AAC32C6F
#define CUSTOMER_H_HEADER_INCLUDED_AAC32C6F
#include "Person.h"
//##ModelId=553C9B4C0295
class Customer : public Person
{
public:
//##ModelId=553C9B890326
Make_Payment();
private:
//##ModelId=553C9B5A003E
Person Name;
//##ModelId=553C9B5E0028
Person Phone;
//##ModelId=553C9B64019E
person Address;
//##ModelId=553C9B680189
String Credit_Card;
//##ModelId=553C9B6F0007
Integer Cash_Amount;
};
```

```
#endif /*
CUSTOMER_H_HEADER_INCLUDED_AAC32C6F */
#ifndef
MARKETING_H_HEADER_INCLUDED_AAC354F0
#define
MARKETING_H_HEADER_INCLUDED_AAC354F0
//##ModelId=553C983603A7
Class Marketing
{
public:
//##ModelId=553C987002CA
Maintain_Catalog();
//##ModelId=553C98770138
Access_Information();
private:
//##ModelId=553C984200B3
String Product_Name;
//##ModelId=553C984C02E5
String Sales_Status;
//##ModelId=553C985E02BD
String Product_Status;
};
#endif /*
MARKETING_H_HEADER_INCLUDED_AAC354F0 */
#ifndef PERSON_H_HEADER_INCLUDED_AAC30F36
#define PERSON_H_HEADER_INCLUDED_AAC30F36
//##ModelId=553C970F00AC
class Person
{
public:
//##ModelId=553C976E01C4
GetDeatils();
```

```
//##ModelId=553C97750038
SetDeatils();
private:
//##ModelId=553C971B02D3
String Name;
//##ModelId=553C97380226
String Phone;
//##ModelId=553C975B0001
String Address;
};
#endif /*
PERSON_H_HEADER_INCLUDED_AAC30F36 */
#ifndef
SALES_ASSISTANT_H_HEADER_INCLUDED_AAC352A
6
#define
SALES_ASSISTANT_H_HEADER_INCLUDED_AAC352A
6
#include "Person.h"
//##ModelId=553C997601E5
class Sales Assistant : public Person
{
public:
//##ModelId=553C9A21007E
Process_Order();
//##ModelId=553C9A260121
Check_Status();
//##ModelId=553C9A300352
Process_Refund();
//##ModelId=553C9A3E007C
Place_Back_Order();
//##ModelId=553C9A4A0191
```

```
Assign_Sales();
//##ModelId=553C9A530207
Collects_Payment();
//##ModelId=553C9A70013A
opname();
private:
//##ModelId=553C99860376
Person Name;
//##ModelId=553C998A0350
Person Phone;
//##ModelId=553C998E035F
Person Address;
};
#endif /*
SALES_ASSISTANT_H_HEADER_INCLUDED_AAC352A
6 */
#ifndef
STOCK_MANAGER_H_HEADER_INCLUDED_AAC32E0
A
#define
STOCK_MANAGER_H_HEADER_INCLUDED_AAC32E0
A
#include "Person.h"
//##ModelId=553C99010386
class Stock Manager : public Person
{
public:
//##ModelId=553C99340150
Check_Demand();
//##ModelId=553C99380262
Monitor_Stock_Level();
//##ModelId=553C9940011C
```

```
Monitor_Weekly_Run_Rates();
//##ModelId=553C994F005E
Order_Stock();
private:
//##ModelId=553C9912016D
Person Name;
//##ModelId=553C991601B4
Person Phone;
//##ModelId=553C991B0229
Person Address;
//##ModelId=553C9920005D
String Stock_Status;
};
#endif /*
STOCK_MANAGER_H_HEADER_INCLUDED_AAC32E0
A */
#ifndef
STORE_MANAGER_H_HEADER_INCLUDED_AAC332A
6
#define
STORE_MANAGER_H_HEADER_INCLUDED_AAC332A
6
#include "Person.h"
//##ModelId=553C97A103B6
class Store Manager : public Person
{
public:
//##ModelId=553C9806018D
Print_Report();
//##ModelId=553C980D0338
Assign_Sales();
//##ModelId=553C981A01B5
```

```
Calculate_Weekly_Sales_Bonus();
private:
//##ModelId=553C97B30348
Person Name;
//##ModelId=553C97BD029E
Person Phone;
//##ModelId=553C97C5007C
Person Address;
//##ModelId=553C97D101AE
String Sales_Status;
//##ModelId=553C97E402A7
String Store_Address;
};
#endif /*
STORE_MANAGER_H_HEADER_INCLUDED_AAC332A6
*/
#ifndef SYSTEM_H_HEADER_INCLUDED_AAC3366C
#define SYSTEM_H_HEADER_INCLUDED_AAC3366C
//##ModelId=553C9A8201AC
class System
{
public:
//##ModelId=553C9ACD031A
Print_Receipt();
//##ModelId=553C9AD20266
Display_Stock_Details();
private:
//##ModelId=553C9AA90341
String Product_Description;
//##ModelId=553C9AB30185
Integer Product_Price;
//##ModelId=553C9ABD00FD
```

```
Integer Stock;
};
#endif /*
 SYSTEM_H_HEADER_INCLUDED_AAC3366C */
#ifndef
WAREHOUSE_STAFF_H_HEADER_INCLUDED_AAC32A
1B
#define
WAREHOUSE_STAFF_H_HEADER_INCLUDED_AAC32A1B
//##ModelId=553C9C2202FC
class WareHouse Staff
{
public:
//##ModelId=553C9C4B0368
Book_Stock();
private:
//##ModelId=553C9C2B02DE
String Stock_Status;
//##ModelId=553C9C330009
Stock_Description;
};
#endif /*
WAREHOUSE_STAFF_H_HEADER_INCLUDED_AAC32
A1B */
#include "Customer.h"
//##ModelId=553C9B890326
Customer::Make_Payment()
{
}
#include "Marketing.h"
//##ModelId=553C987002CA
Marketing::Maintain_Catalog()
```

```
{
}
//##ModelId=553C98770138
Marketing::Access_Information()
{
}
#include "Person.h"
//##ModelId=553C976E01C4
Person::GetDeatils()
{
}
//##ModelId=553C97750038
Person::SetDeatils()
{
}
#include "Sales Assistant.h"
//##ModelId=553C9A21007E
Sales Assistant::Process_Order()
{
}
//##ModelId=553C9A260121
Sales Assistant::Check_Status()
{
}
//##ModelId=553C9A300352
Sales Assistant::Process_Refund()
{
}
//##ModelId=553C9A3E007C
Sales Assistant::Place_Back_Order()
{
}
```

```
//##ModelId=553C9A4A0191
Sales Assistant::Assign_Sales()
{
}
//##ModelId=553C9A530207
Sales Assistant::Collects_Payment()
{
}
//##ModelId=553C9A70013A
Sales Assistant::opname()
{
}
#include "Stock Manager.h"
//##ModelId=553C99340150
Stock Manager::Check_Demand()
{
}
//##ModelId=553C99380262
Stock Manager::Monitor_Stock_Level()
{
}
//##ModelId=553C9940011C
Stock Manager::Monitor_Weekly_Run_Rates()
{
}
//##ModelId=553C994F005E
Stock Manager::Order_Stock()
{
}
#include "Store Manager.h"
//##ModelId=553C9806018D
Store Manager::Print_Report()
```

```
{
}
//##ModelId=553C980D0338
Store Manager::Assign_Sales()
{
}
//##ModelId=553C981A01B5
Store Manager::Calculate_Weekly_Sales_Bonus()
{
}
#include "System.h"
//##ModelId=553C9ACD031A
System::Print_Receipt()
{
}
//##ModelId=553C9AD20266
System::Display_Stock_Details()
{
}
#include "WareHouse Staff.h"
//##ModelId=553C9C4B0368
WareHouse Staff::Book_Stock()
{
}
```

This process of code generation is also we can say implementation of forward engineering in C++ for store stock control based web application. Further, through roundtrip engineering, we develop the understanding level of complex software. In the above experimental case, we have transformed the object oriented models into C++ source code. With the help of above software development process, developers can identify software Metrics like no. of data member per class, friend functions and friend classes, data structure metrics, coupling between objects, information flow etc. We can calculate the size of the vocabulary of a program, no of unique operators, no of unique operands, occurrences of operators and operands, the length of the program, Volume, potential Volume, Program Level, Program Difficulty, Effort, average number of live variables, Count of executable statements and total number of variables etc. with the help of above source code software project planning will become more and more easy.

This forward engineering process for software development will also help the developer about the cohesion and coupling. As we know, Cohesion means, what is a class's (or module's) purpose? Low cohesion implies that a class does a variety of actions and isn't focused on anything particular. On the contrary high cohesion means that a class is focused on what exactly it should be doing. The transition process of forward and reverse engineering is as follows:- As we know, Round-trip engineering is the capability to create object oriented UML model from source code and generate source code from these object oriented UML model, and keep them synchronized. For any web application, we can use round-trip engineering to remain our implementation model and source code up-to-date. Object oriented design information is improved from the source code and some obtainable design documentation.

The procedure of improving a program's design is known as design recovery. A design is improved by piecing equally information from the source code, obtainable documents, software developer's experienced with the software system knowledge. As we know, the software roundtrip engineering i.e. forward engineering and backward engineering (as we can notice in Fig. 4.1.6 and Fig. 4.1.7) plays a vital role in software development life cycle. In this book, we have used round trip engineering implementation for the same stock control system web application.

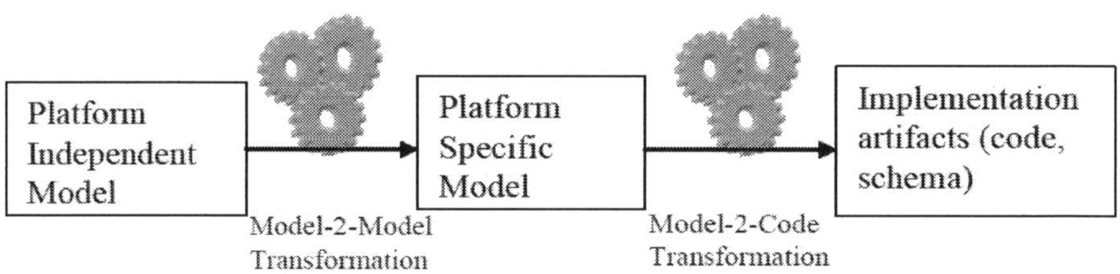

Fig. 4.1.6: Transformation approach in forward engineering.

Software's forward engineering and backward engineering, we can notice in Fig. 4.1.7 and this round trip engineering plays a vital role in software development life cycle. Fig. 4.1.7 demonstrates the transformation approach in reverse engineering.

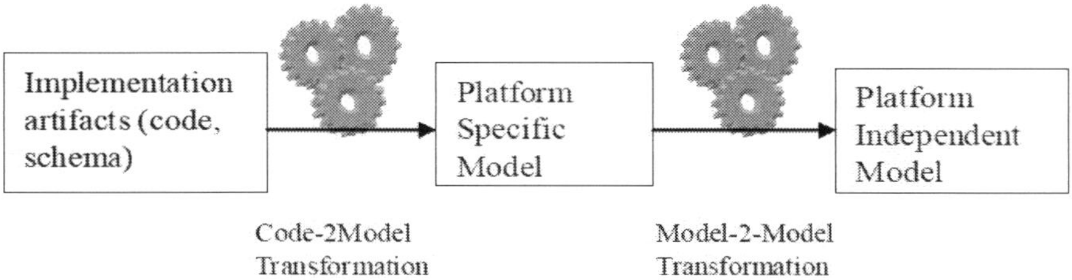

Fig. 4.1.7: Transformation approach in reverse engineering.

As we know, in Reverse Engineering, we study the complete system to identify all the components of the system and their interrelationships so as to ascertain which components can be reused, and how they can be reused. Fig. 4.1.8 indicates the same software development process. Forward engineering starts with problem statements and reverse engineering start with existing software with source code. In order to create a representation of the system in another form or at higher level of abstraction, reverse engineering is must. Reverse engineering is used to increase the understanding level about the software. In this process, SRS, SDD any diagram or any document can be used, so that maintenance engineer can understand the existing software for the purpose of up gradation. Fig. 4.1.9 shows the implementation model of Software Round trip engineering process for complex software development.

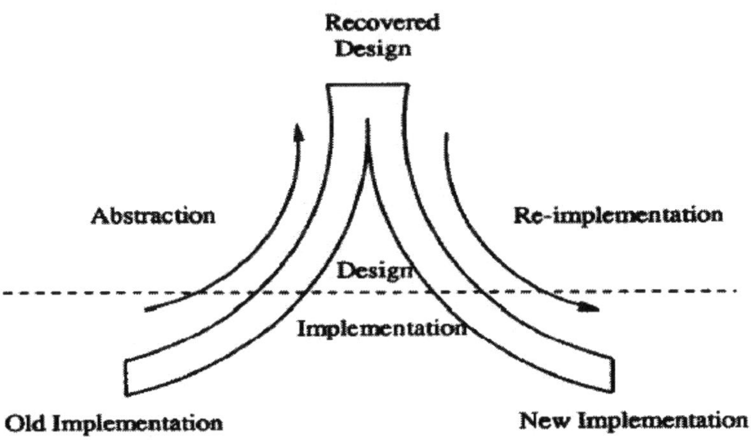

Fig. 4.1.8: Reverse engineering and re-implementation process for web applications

There are various business object oriented tools that provide the reverse engineering abilities. In this book we have used IBM Rational Rose for the same purpose. The information can furthermore be used to develop and preserve other software systems; i.e. we achieve supplementary information that can be used by a forward engineer for the purpose of forward engineering.

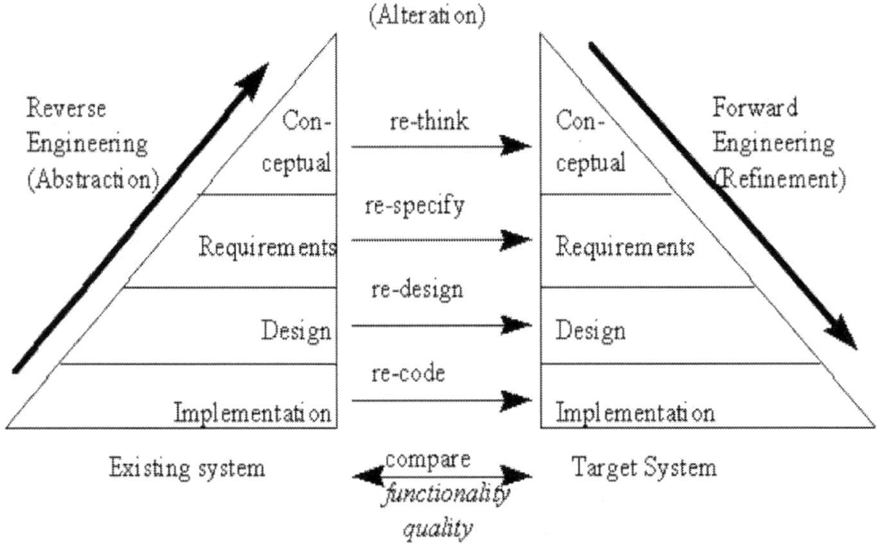

Fig. 4.1.9: General model for software round trip engineering

UML tools mostly apply reverse engineering to execute round-trip engineering. They convert back source code into a model. As we know the software does not wear out. The concept of Reverse engineering has become a major concern in software business because of their complete size and complexity. The most creative way to software reverse engineer a database is to build an object oriented model that conveys the software's scope and software's objective. Evaluate the generally software quality of the software system and increase an impression of the software system in terms of software size, software complexity, and software structure. Locate and recognize the mainly significant object oriented classes and find the object oriented classes and hierarchies that indicate a core division of the software system's domain and recognize their software structure in terms of implementation and purpose in terms of software functionality.

As we know, it is very easy to modify an object oriented design as compared to its source code. The recovered design of old software describes the existing software system, after that we can design and develop more secure system. After reverse engineering and roundtrip engineering of any old web application, we will get a new software system that is well structured, properly documented, and easier to maintain as compared to the old and previous software version.

4.2 OBJECT ORIENTED SECURE ROUND TRIP ENGINEERING FOR WEB APPLICATIONS-2

In this book we are taking one more case study and this case study is related to secure product purchase system. We propose object oriented software development process to design and develop the above case study. Consider the following functional and non functional requirements for above case study:-

- The administrator of this web application updates the inventory stock. Every time inventory stock is updated by administrator, inventory data is saved to a disk of the devise.
- Generalization of this application can be of two specialized kinds: (1) make-a phone order sale; and (2) make-a walk-in sale. Generalization shows the "kind of "or "is a" relationship. In object oriented language this concept implemented through inheritance.
- A sales clerk of this software portal records the make-a-walk-in sales.
- Every day, whenever there is a sale of product, the stock of inventory is updated.
- A sale may require validating a credit card if the purchase is through credit card option. Same verification and authentication will take place if payment is through debit card and net banking. During implementation we have to take care all these consideration.

The anticipated approach is based on the object-oriented approach. This approach has the improvement mechanisms over procedural language. Object oriented analysis and designing approach is better than structure analysis and designing due to real world mapping of problems. This OOA and OOD make our development process more simple and realistic. The major phases of the software development process are as follows:

- Mine applications requirements through requirement discovery and requirement elicitation
- Develop the use case diagram (UCD) according to user's requirement for specific web application. This diagram will vary from one application to another
- Indicate every use case in a textual format i.e. description of use case to develop the understanding level of the software
- Demonstrate dynamic behavior with the help of activity diagrams for specific web applications
- Convert these object oriented models into a class diagram (CD) for further forward engineering
- Forward engineering i.e. code generation of this class diagram in object oriented programming language like C++ and Java etc.

First of all we have to identify the functional requirement in any system. After the requirement analysis for any software, we demonstrate the major functional requirements through use case diagram. Designing the use case diagram for any system is essential because 'Use Cases' are a prominent way for the requirement elicitation and documentation. Requirement elicitation and documentation will be completed through software requirement specifications (SRS).Use cases of any application show the functional requirement and services that are requested and desired by customer. Use cases are explained in natural languages so that every analyst, designer, coder can easily understand the real and realistic requirement of customer.

Various requirements discovery approaches are there to implement like group discussion, felicitated application specification techniques (FAST), Quality functional deployment (QFD), interview, Delphi method and form designing etc. With these approaches pattern of functions or methods can be identified in any application for implementation. Use case approach for object orientation is natural implementation approach to understand and present an excellent way for communicating with clients and users. For secure web application development, Use case based approach support and manages the complexity of huge and complex projects. Complexity of huge and complex projects can be minimized by decomposing the complexity into most important functions (i.e., use cases or services that the software will provide to customer). Fig. 4.2.1 shows the generalized use case diagram for above case study.

Fig. 4.2.1 demonstrate various functional requirement in given case study, these requirements represented by use case like save, load, run inventory report, update , make a sale, make phone order sale and purchase etc. it means projected system after deployment will provide these services to consumers. According to Fig. 4.2.1, we can write description of major Use cases for the understanding of web application. This understanding level can be used during the phases of software development life cycle. These use case description can help in analysis, designing, coding and maintenance etc. Tables 4.2.1 through 4.2.11 show the description of major use cases of online product purchase based web Application.

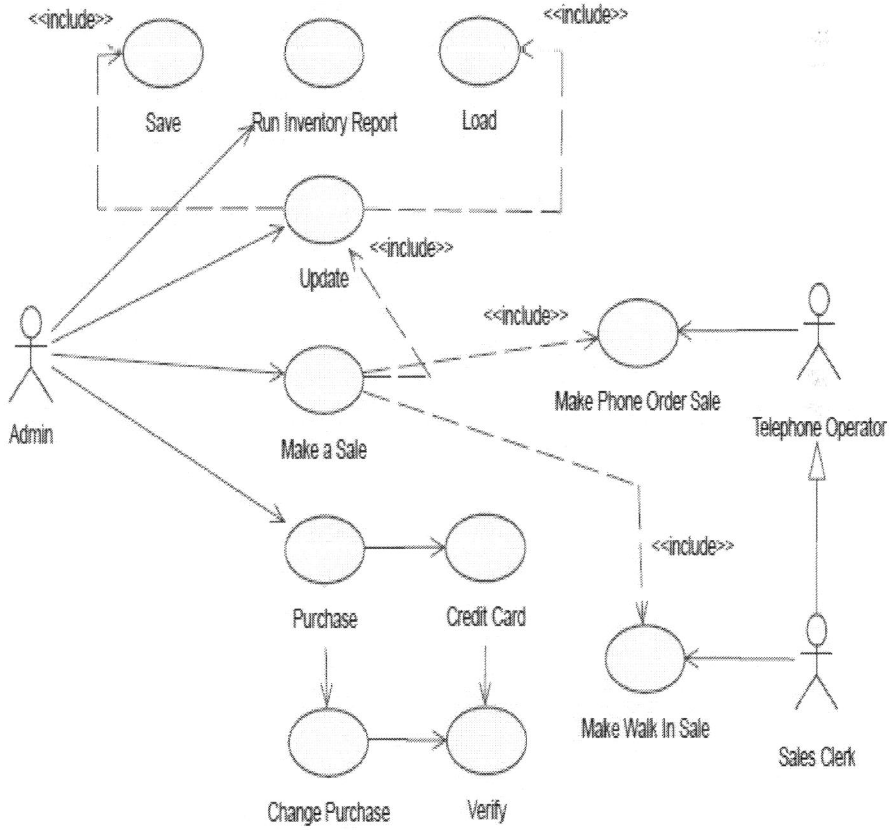

Fig. 4.2.1: Standard Use case diagram for online product purchase application

Table 4.2.1: Run inventory report use case description

Unique ID of Use Case	**1**
Name of the Use Case	Run Inventory Report
Objective	This is done for getting information about the available inventory
Actors	Administrator
Pre-Conditions	Inventory information mus be available on the disk
Post-conditions	Inventory report is provided to the the concerned actor
Flow of Events of Run inventory report use case	This Run Inventory Report use case starts when the administrator wishes to get the inventory report, this initiates the work of loading information from the disk and then report is generated
Alternative Flow	None

Table 4.2.2: Use case description of update functional requirement

Use Case	**2**
Use Case Name	Update
Objective	This is done for updating information regarding inventory
Actors	Administrator
Pre-Conditions	Inventory information mus be available on the disk to be updated
Post-conditions	Updation is done successfully
Flow of Events of update use case	This update use case starts when the administrator wishes to update the inventory report, this initiates the work of loading information from the disk and then saving it again on the disk after making changes to it.
Alternative Flow	None

Table 4.2.3: Use case description of load use case

Unique ID of Use Case	**3**
Use Case Name	Loads
Objective	This is done for getting data from the disk

Unique ID of Use Case	3
Actors	None
Pre-Conditions	Data must be available in the disk
Post-conditions	Data is loaded from the disk
Flow of Events of load use case	This load use case starts when the one wishes to have the inventory information, this initiates the work of loading information from the disk
Alternative Flow	None

Table 4.2.4: Use case description of save report

Use Case	4
Use Case Name	save
Objective	This is done for saving information about the available inventory
Actors	None
Pre-Conditions	None
Post-conditions	Inventory information is saved on the disk
Flow of Events	The execution of this use case take place when one wishes to save the inventory information , this initiates by making changes to pre-loaded data and saving it back to the disk
Alternative Flow	None

Table 4.2.5: Use case description of make a sale

Use Case	5
Use Case Name	Make a sale
Objective	This comes into picture when there is sale for some product
Actors	Administrator
Pre-Conditions	Product must be available for the sale
Post-conditions	Product is sold and regarding changes are reflected to the data on the disk
Flow of Events	The execution of this use case take place when the one wishes to sell a product , when this happens use case “update” does the work for making changes after selling the product
Alternative Flow	None

Table 4.2.6: Use case description of make phone order sale

Use Case	6
Use Case Name	Make phone order sale
Objective	This is done for selling product by taking order on the phone
Actors	Telephone Operator
Pre-Conditions	Product must be available for the sale
Post-conditions	Product is sold and regarding changes are reflected to the data on the disk
Flow of Events	The execution of this use case take place when the one wishes to sell a product , when this happens use case “update” does the work for making changes after selling the product
Alternative Flow	None

Table 4.2.7: Use case description of make walkin sale use case

Use Case	7
Name of Use Case	Make walkin sale
Objective	This is done for selling product
Actors	Sales Clerk
Pre-Conditions	Product must be available for the sale
Post-conditions	Product is sold and regarding changes are reflected to the data on the disk
Flow of Events	The execution of this use case take place when the one wishes to sell a product , when this happens use case “update” does the work for making changes after selling the product
Alternative Flow	None

Table 4.2.8 Use case description of purchase use case

Unique ID of Use Case	8
Name of Use Case	Purchase
Objective	This is used when some product is purchased
Actors	Administrator
Pre-Conditions	Product must be available for the purchase
Post-conditions	If payment made successfully product is purchased else not

Unique ID of Use Case	8
Flow of Events	The execution of this use case take place when the one wishes to purchase a product , when this happens payment is asked for and if everything went on successfully then product is purchased
Alternative Flow	Product is not purchased

Table 4.2.9: Use case description of credit card

Use Case	9
Use Case Name	Credit card
Objective	this is used when chosen mode of payment is credit card
Actors	None
Pre-Conditions	Credit card must be available with the purchaser
Post-conditions	Payment done via credit card
Flow of Events	The execution of this use case take place when the one wishes to pay by credit card if all the rules are followed then payment is done
Alternative Flow	Payment not done

Table 4.2.10: Use case description of cheque purchase

Use Case	10
Use Case Name	Cheque Purchase
Objective	this is used when chosen mode of payment is cheque
Actors	None
Pre-Conditions	Cheque must be available with the purchaser
Post-conditions	Payment done via cheque
Flow of Events	The execution of this use case take place when the one wishes to pay by cheque if all the rules are followed then payment is done
Alternative Flow	Payment not done

Table 4.2.11: Use case description of verifies

Unique ID of Use Case	11
Name of Use Case	Verifies
Objective	this is used when either payment mode is to be verified against the predefined norms

Unique ID of Use Case	**11**
Actors	None
Pre-Conditions	None
Post-conditions	Conditions are checked for and status is returned to parent use-case
Flow of Events	The execution of this use case take place when the one wishes verify the payment mode against predefined norms. Checks them and returns the status
Alternative Flow	None

The mainly universal approach to use UML diagrams during analysis is to initiate with use case diagrams for capturing the functional requirement of user. This result can be complemented by designing an activity diagram for the same case study. Activity diagram demonstrates how use cases or functional requirements are related to each other within the application. Fig. 4.2.2 shows the well accepted activity diagram for product purchase web application. The activity diagram illustrates the various activities and overall flow of control to describe a whole software system. The common activity diagram of above case study is as follows:–

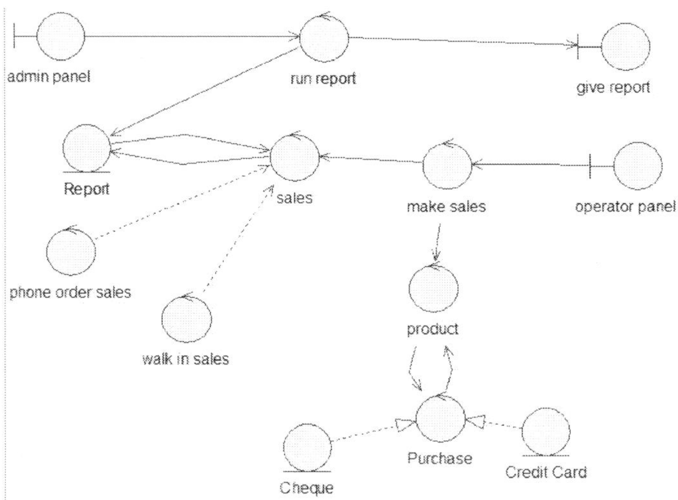

Fig. 4.2.3. Analysis model for product purchase based system

Fig. 4.2.3 shows the analysis model for product purchase web application. This analysis model covers all interface object of product purchase, control object and entity object of product purchase. In Fig.

4.2.3 shows admin panel, give report and operator panel as interface object in product purchase system. Run report, sales, make sales, product, phone order sales and purchase etc. are the control objects of product purchase system. Credit card, cheque and report are the example of entity objects in above case study.

The implementation concept into uml 2.0 is possible through IBM Rational Rose. Object oriented Software modularization, decomposition in modules, is a system approach for improving the structure and understanding of source code. Every language uses the concept of modularity. Procedures, functions, objects and modules are the example of modularity. In order to understand software, software developers need to make a well-connected demonstration of the classes that make up the desired system. Each and every class must be understood individually and, then, relationships among classes as well. Further data members of class, member functions of the class, friend function and visibility level etc. are also major concern during the designing of class diagram for system. Development and implementation of any software will be up to the mark if analysis and design is perfect in all respect. So analysis is close to design and design should be close to implementation. Decomposition of problem statement into meaningful classes in an attempt to control the complexity of large and complex software.

At analysis phase, analyst designs the SRS and draws the use case diagram for the same case study. For performing forward engineering or code generation in C++, with the help of IBM Rational Rose Software, we design and develop the class diagram for the same web application [2][15]. Class diagrams are the most and widely used Unified modelling language diagram in the community of developers and coder. Concepts of classes are very helpful in the software development process from the initial phase of SDLC. In UML, a class is represented as a box with three vertical sections. The top section of class shows the name of the class. The middle section displays the variables or data members with visibility level (public, protected, or private) and properties (constant or static). The bottom section contains the member functions or member methods of the particular class.

The class diagram for online product purchase based application is feasible, which shows various storing objects like Credit card, sales clerk, Payment, Person, Admin and Product, etc. (see Fig. 4.2.4). The properties i.e. data members and member functions will vary from one application to another. Class wise code, or source code in C++ is given below; this C++ source code can help the maintenance engineer, designers, and coders during the development and maintenance process of software. Fig. 4.2.4 shows the class wise resultant class diagram and dummy code or blue print in C++ for further object oriented software development.

Forward engineering and code generation in C++ source Code for online product purchase system is as follows:

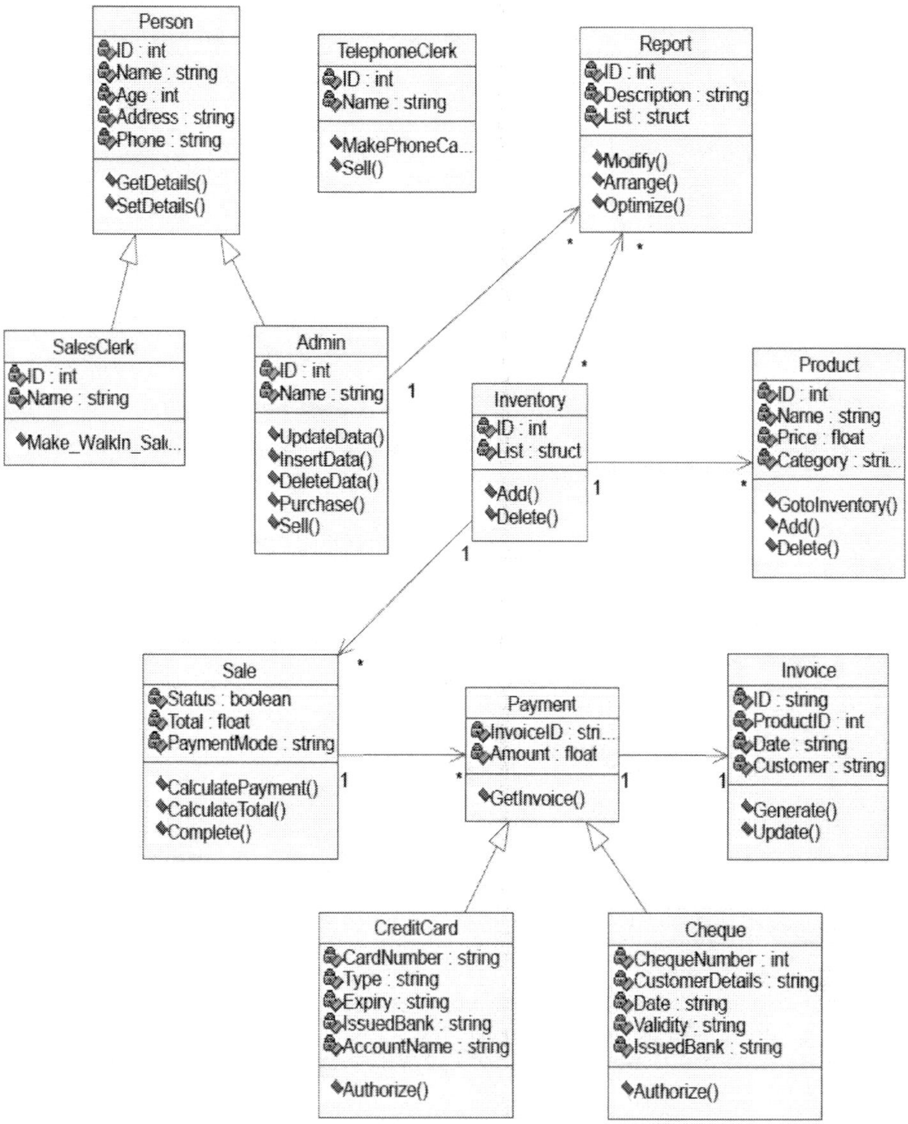

Fig. 4.2.4: Standard class diagram for product purchase application

```
#include "Admin.h"
//##ModelId=5469AC6603B6
Boolean Admin::UpdateData()
{
}
//##ModelId=5469AC6F001F
Boolean Admin::InsertData()
{
}
//##ModelId=5469AC73022B
Boolean Admin::DeleteData()
{
}
//##ModelId=5469AC7800CC
Admin::Purchase()
{
}
//##ModelId=5469AC7B02F9
Admin::Sell()
{
}
#ifndef ADMIN_H_HEADER_INCLUDED_AB9602D6
#define ADMIN_H_HEADER_INCLUDED_AB9602D6
#include "Person.h"
//##ModelId=5469AC3D0259
class Admin : public Person
{
  public:
    //##ModelId=5469AC6603B6
    Boolean UpdateData();
    //##ModelId=5469AC6F001F
    Boolean InsertData();
```

```
    //##ModelId=5469AC73022B
    Boolean DeleteData();
    //##ModelId=5469AC7800CC
    Purchase();
    //##ModelId=5469AC7B02F9
    Sell();
  private:
    //##ModelId=5469AC4500F5
    int ID;
    //##ModelId=5469AC480322
    string Name;
};
#endif /* ADMIN_H_HEADER_INCLUDED_AB9602D6 */
#include "Cheque.h"
//##ModelId=5469B05C0153
Cheque::Authorize()
{
}
#ifndef CHEQUE_H_HEADER_INCLUDED_AB962A29
#define CHEQUE_H_HEADER_INCLUDED_AB962A29
#include "Payment.h"
//##ModelId=5469B01C02E2
class Cheque : public Payment
{
  public:
    //##ModelId=5469B05C0153
    Authorize();
  private:
    //##ModelId=5469B0280001
    int ChequeNumber;
    //##ModelId=5469B0300300
    string CustomerDetails;
```

```
    //##ModelId=5469B03D0108
    string Date;
    //##ModelId=5469B0460147
    string Validity;
    //##ModelId=5469B04C03DD
    string IssuedBank;
};
#endif /* CHEQUE_H_HEADER_INCLUDED_AB962A29 */
#include "CreditCard.h"
//##ModelId=5469B00D0376
CreditCard::Authorize()
{
}
#ifndef CREDITCARD_H_HEADER_INCLUDED_AB965C60
#define CREDITCARD_H_HEADER_INCLUDED_AB965C60
#include "Payment.h"
//##ModelId=5469AFAE03D0
class CreditCard : public Payment
{
  public:
    //##ModelId=5469B00D0376
    Authorize();
  private:
    //##ModelId=5469AFC30312
    string CardNumber;
    //##ModelId=5469AFD200D8
    string Type;
    //##ModelId=5469AFDA0010
    string Expiry;
    //##ModelId=5469AFE5002E
    string IssuedBank;
    //##ModelId=5469AFF1003A
```

```
    string AccountName;
};
#endif /* CREDITCARD_H_HEADER_INCLUDED_AB965C60 */
#include "Inventory.h"
//##ModelId=5469AD9E032B
Inventory::Add()
{
}
//##ModelId=5469ADA103C2
Inventory::Delete()
{
}
#ifndef INVENTORY_H_HEADER_INCLUDED_AB961BCB
#define INVENTORY_H_HEADER_INCLUDED_AB961BCB
//##ModelId=5469AD7F01F5
class Inventory
{
  public:
    //##ModelId=5469AD9E032B
    Add();
    //##ModelId=5469ADA103C2
    Delete();
  private:
    //##ModelId=5469AD910246
    int ID;
    //##ModelId=5469AD950368
    struct List;
};
#endif /* INVENTORY_H_HEADER_INCLUDED_AB961BCB */
#include "Invoice.h"
```

```
//##ModelId=5469AF7E0007
Invoice::Generate()
{
}
//##ModelId=5469AF820129
Invoice::Update()
{
}
#ifndef INVOICE_H_HEADER_INCLUDED_AB960652
#define INVOICE_H_HEADER_INCLUDED_AB960652
//##ModelId=5469AF360210
class Invoice
{
 public:
  //##ModelId=5469AF7E0007
  Generate();
  //##ModelId=5469AF820129
  Update();
 private:
  //##ModelId=5469AF410069
  string ID;
  //##ModelId=5469AF460254
  int ProductID;
  //##ModelId=5469AF570352
  string Date;
  //##ModelId=5469AF610017
  string Customer;
};
#endif /* INVOICE_H_HEADER_INCLUDED_AB960652 */
#include "Payment.h"
//##ModelId=5469AEF00203
Payment::GetInvoice()
```

```
{
}
#ifndef PAYMENT_H_HEADER_INCLUDED_AB966288
#define PAYMENT_H_HEADER_INCLUDED_AB966288
//##ModelId=5469AECB01F0
class Payment
{
  public:
    //##ModelId=5469AEF00203
    GetInvoice();
  private:
    //##ModelId=5469AED703E0
    string InvoiceID;
    //##ModelId=5469AEE30266
    float Amount;
};
#endif /* PAYMENT_H_HEADER_INCLUDED_AB966288 */
#include "Person.h"
//##ModelId=5469ABC1016B
Boolean Person::GetDetails()
{
}
//##ModelId=5469ABCE0195
Boolean Person::SetDetails()
{
}
#ifndef PERSON_H_HEADER_INCLUDED_AB9669C7
#define PERSON_H_HEADER_INCLUDED_AB9669C7
//##ModelId=5469AB300078
class Person
{
  public:
```

```
    //##ModelId=5469ABC1016B
    Boolean GetDetails();
    //##ModelId=5469ABCE0195
    Boolean SetDetails();
  private:
    //##ModelId=5469AB700122
    int ID;
    //##ModelId=5469AB7802AB
    string Name;
    //##ModelId=5469AB810184
    int Age;
    //##ModelId=5469AB8E0140
    string Address;
    //##ModelId=5469AB9403E5
    string Phone;
};
#endif /* PERSON_H_HEADER_INCLUDED_AB9669C7 */
#include "Product.h"
//##ModelId=5469ADE600B6
Product::GotoInventory()
{
}
//##ModelId=5469ADED034A
Product::Add()
{
}
//##ModelId=5469ADEF02BA
Product::Delete()
{
}
#ifndef PRODUCT_H_HEADER_INCLUDED_AB967F99
#define PRODUCT_H_HEADER_INCLUDED_AB967F99
```

```
//##ModelId=5469ADB50078
class Product
{
  public:
    //##ModelId=5469ADE600B6
    GotoInventory();
    //##ModelId=5469ADED034A
    Add();
    //##ModelId=5469ADEF02BA
    Delete();
  private:
    //##ModelId=5469ADC2018C
    int ID;
    //##ModelId=5469ADC603B7
    string Name;
    //##ModelId=5469ADCC0005
    float Price;
    //##ModelId=5469ADD2000B
    string Category;
};
#endif /* PRODUCT_H_HEADER_INCLUDED_AB967F99 */
#include "Report.h"
//##ModelId=5469ACD8017A
Report::Modify()
{
}
//##ModelId=5469ACDC000D
Report::Arrange()
{
}
//##ModelId=5469ACDF03B0
Report::Optimize()
```

```
{
}
#ifndef REPORT_H_HEADER_INCLUDED_AB967D81
#define REPORT_H_HEADER_INCLUDED_AB967D81
//##ModelId=5469ACB4029E
class Report
{
  public:
    //##ModelId=5469ACD8017A
    Modify();
    //##ModelId=5469ACDC000D
    Arrange();
   //##ModelId=5469ACDF03B0
    Optimize();
  private:
    //##ModelId=5469ACBD0119
    int ID;
    //##ModelId=5469ACC10392
    string Description;
    //##ModelId=5469ACCA0114
    struct List;
};
#endif /* REPORT_H_HEADER_INCLUDED_AB967D81 */
#include "Sale.h"
//##ModelId=5469AEA7002D
Sale::CalculatePayment()
{
}
//##ModelId=5469AEAE00AF
Sale::CalculateTotal()
{
}
```

```
//##ModelId=5469AEB70081
Sale::Complete()
{
}
#ifndef SALE_H_HEADER_INCLUDED_AB966A8D
#define SALE_H_HEADER_INCLUDED_AB966A8D
//##ModelId=5469AE7703E4
class Sale
{
  public:
    //##ModelId=5469AEA7002D
    CalculatePayment();
    //##ModelId=5469AEAE00AF
    CalculateTotal();
    //##ModelId=5469AEB70081
    Complete();
  private:
    //##ModelId=5469AE870013
    boolean Status;
    //##ModelId=5469AE8E00E2
    float Total;
    //##ModelId=5469AE95000C
    string PaymentMode;
};
#endif /* SALE_H_HEADER_INCLUDED_AB966A8D */
#include "SalesClerk.h"
//##ModelId=5469AC0B00D6
SalesClerk::Make_WalkIn_Sale()
{
}
#ifndef SALESCLERK_H_HEADER_INCLUDED_AB966F1D
#define SALESCLERK_H_HEADER_INCLUDED_AB966F1D
```

```
#include "Person.h"
//##ModelId=5469AC000329
class SalesClerk : public Person
{
  public:
    //##ModelId=5469AC0B00D6
    Make_WalkIn_Sale();
  private:
    //##ModelId=5469AC250157
    int ID;
    //##ModelId=5469AC2B0025
    string Name;
};
#endif /* SALESCLERK_H_HEADER_INCLUDED_AB966F1D */
#include "TelephoneClerk.h"
//##ModelId=5469AE4400EB
TelephoneClerk::MakePhoneCall()
{
}
//##ModelId=5469AE4A0064
TelephoneClerk::Sell()
{
}
#ifndef TELEPHONECLERK_H_HEADER_INCLUDED_AB964331
#define TELEPHONECLERK_H_HEADER_INCLUDED_AB964331
//##ModelId=5469AE2B03B2
class TelephoneClerk
{
  public:
    //##ModelId=5469AE4400EB
    MakePhoneCall();
    //##ModelId=5469AE4A0064
```

```
    Sell();
    private:
      //##ModelId=5469AE38014C
      int ID;
      //##ModelId=5469AE3B0231
      string Name;
  };
  #endif /* TELEPHONECLERK_H_HEADER_INCLUDED_AB964331 */
```

In this proposed scheme, during the forward engineering process, here we have converted the OO models into the OO programming languages like Java, C++ and Visual Basic etc. In this case study we implemented the forward engineering to generate the source code in C++ programming language. With the help of this C++ source code any company can complete the software project planning. Software company can identify the unique operators, unique operands, occurrence of unique operands and operators, volume of the program, potential volume etc. Another interest lies in the mining of relevant information from the source code. We can extract all coding pattern, identifiers and comments from the source code.

4.3 OBJECT ORIENTED SECURE ROUND TRIP ENGINEERING FOR WEB APPLICATIONS-3

We suggest Object Oriented designing for the development of secure applications. During object oriented designing, we can use IBM rational rose for UML modeling purpose. UML tools generate program language source code from UML models. UML can be used for modeling and designing the entire computer e mail system or any other software. Here in this section of book, we are considering the case study of generic Computer e mail. The implementation of computer mailing system with security and performance is possible through object oriented concept.

On the basis of requirement elicitation techniques like FAST, interview, QFD, Questionnaire and survey etc., a system analyst can mine or identify the realistic pattern of user requests in desired software. This requirement mining is possible to any web applications like online shopping, job portals, computer e-mail system, product purchase portals, social networking sites etc. We can identify the pattern like Hit rate, Max operations performed by customers, Frequency of like, post, write, add, share and remark etc. in required application. After identifying the answer of above questions related to user's requirement and pattern of system calls, according to our desired requirement we can disable few protections, security checks in working software to enhance the overall performance of web applications.

A network refers to entities connected together to facilitate communication between them. One of the most widespread examples of a network is World Wide Web, where nodes represent the web pages on the web and edges signify the links through which these pages are connected. A sub-network present on the World Wide Web is a social network. A Social network can be viewed as a network involving people and connections between them. It comprises of a set of actors and ties where the former represents the people in the network and the latter shows the relationships between them [2]. Use of social media has aided such networks tremendously in communicating with their peers openly on the web using words and codes, only they understand. Such networks post regular updates using high quality videos about their beheading activities under different hash tags on twitter which tend to increase their page rank. Use of inspirational and motivational language on twitter attracts a lot of youngsters to join in the mission of these people.

In the SPF based development process does not mean that all UML diagrams must be used and designed. The decision is usually based on the nature, functional and non functional and other requirement of the system. In this section of book, we are considering the implementation of SPF and code generation for the case study of Computer Email System. For this case study first of all we have to design the use case diagram for the representation of expected functional requirement of the system. These functional requirements are services which are expected by user from the system. Functional requirement of computer e mail system can be directly mapped to the Use Cases through UML use case diagram [19]. At analysis phase of software development life cycle, use cases are written in simple language by software analyst. It means use case description for major use cases. Use cases should be simple, so that every designer, developer, tester and maintenance engineer can easily understand the real requirements of user. Use cases can help and minimize the complexity of complex projects, by decomposing the complexity into most considerable functions (i.e., use cases) and by indicating web applications from the users' viewpoint. Fig. 4.3.1 shows the standard Use case diagram for a Computer Email System:-

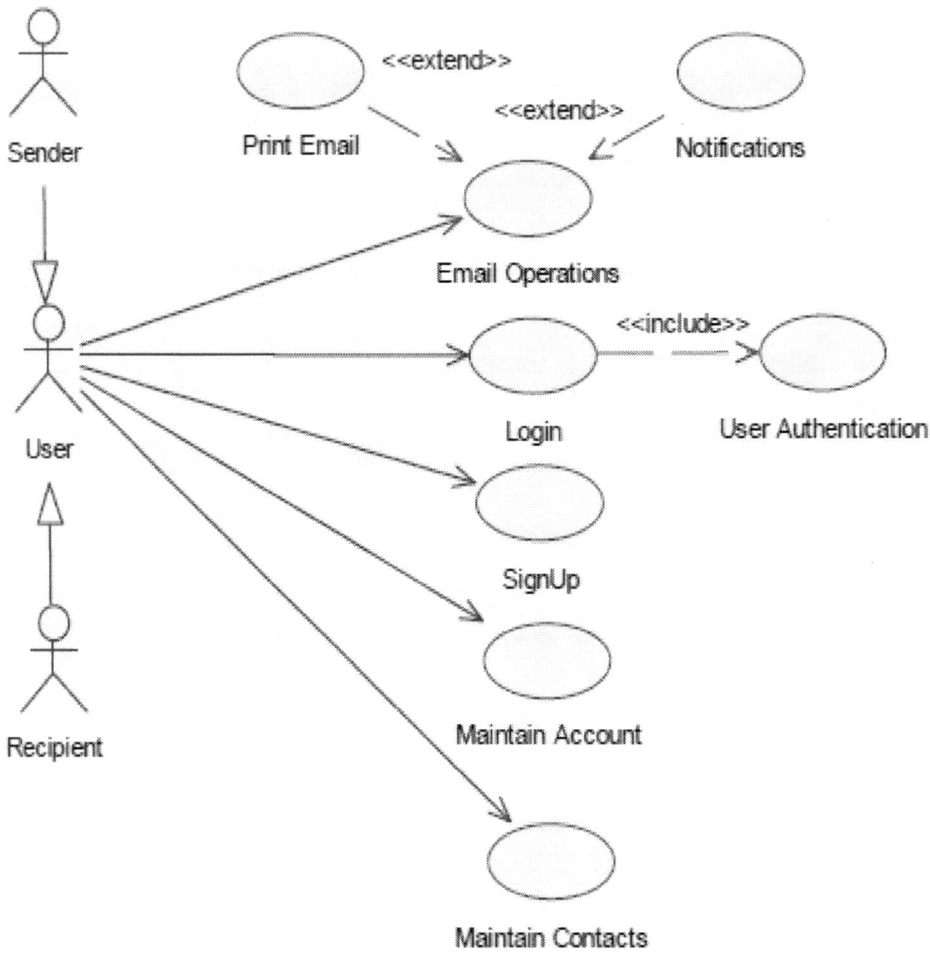

Fig. 4.3.1: Use case diagram for a computer email system

According to Fig. 4.3.1, we can write use case description of major Use cases for the understanding of web application. This understanding level can be used during the phases of software development life cycle. These use case description can help in analysis, designing, coding and maintenance etc. Tables 4.3.1 through 4.3.6 show the description of major use cases of online product purchase based web Application.

Table 4.3.1: Use case description of login use case

Login use case Brief Description	**Explains how a user login to the Email System**
Actors	Sender, Receiver
Flow of Events	• Enter user id • Enter password • Login
Alternative Flow	None
Precondition	Sender/Receiver must have an account
Post condition for Login use case	If the use case executed successfully, the actor is logged in to the system. If not, than the state of the system remain unchanged.

Table 4.3.2 Description of send email use case

Brief Description	**This enables the sender to send an email**
Actors	Sender
Flow of Events	• Log in • Write mail • Send mail
Alternative Flow	None
Precondition	Sender must log in
Post condition	Save sent mail

Table 4.3.3: Description of get email use case

Brief Description	**This use case enables the receiver to receive the email**
Actors	Receiver
Flow of Events	• Log in • Check email • Receive mail
Alternative Flow	None
Precondition	Receiver must login Sender must send mail
Post condition	Save mail/ Delete mail/ Reply

Table 4.3.4: Description of add contacts use case

Brief Description	**This use case enables the receiver/sender to add new contacts**
Actors	Receiver, Sender
Flow of Events	• Log in • Add Name • Add email id
Alternative Flow	If contact already exist, modify
Precondition	Login
Post condition	Save contact

Table 4.3.5: Description of log out use case

Brief Description	**This use case enables the receiver/sender log out from the system after use**
Actors	Receiver, Sender
Flow of Events	• Log in • Add contact/ send/receive email • Log out
Alternative Flow	None
Precondition	Login
Post condition	Close system

Table 4.3.6: Description of print use case

Brief Description	**Use case extended by Get Email, used to print the mail**
Actors	Receiver
Flow of Events	• Log in • Get mail • Print mail
Alternative Flow	None
Precondition	Get mail
Post condition	None

As we know every object has SBI property. SBI stands for state of an object, behavior of an object and identity of an object. Every object has its own state, behavior and identity. Every object has its starting and exit state. This dynamic behavior of any object, we can demonstrate through state transition diagram. As we can see in Fig. 4.3.2 shows the state transition diagram of the Computer Email System web application:–

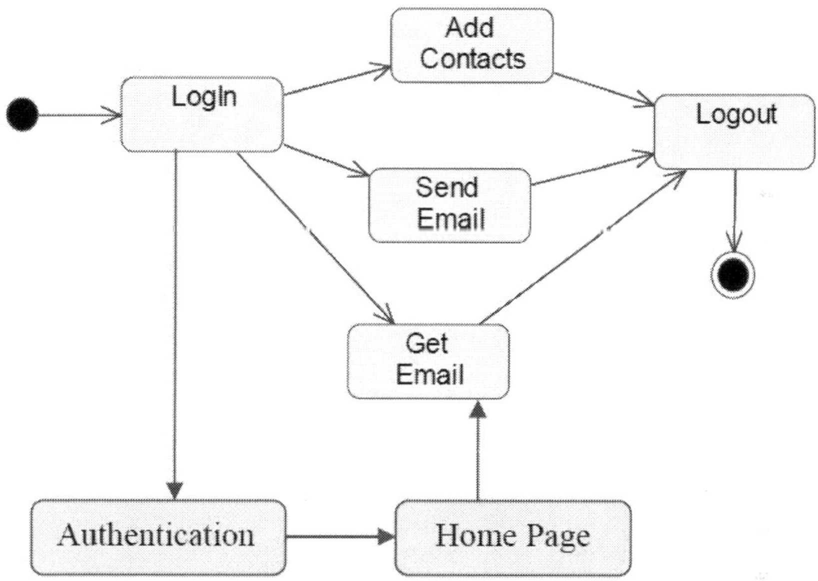

Fig. 4.3.2: State transition diagram showing states which implement the email system

For the above computer e-mail web application, we can design the component class diagram with the assist of above use case diagram (shown in Fig. 4.3.1) and can complete forward engineering for the same computer e mail system through IBM Rational Rose Software. During the SDLC process, Class diagrams are the mainly acknowledged UML diagrams used by each and every software developer. Object oriented class diagram give explanations about the objects and their relationships within the desired computer system. On the basis of code and design, we can implement the SPF model for trusted operating system i.e. well suited for various web applications. Fig. 4.3.3 shows the generic Class Diagram of the Computer Email System web application:—

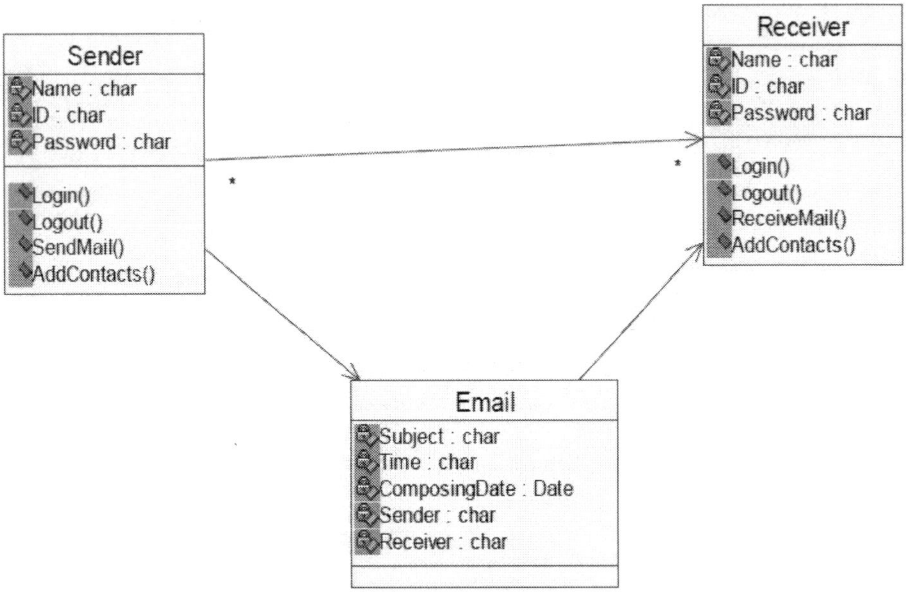

Fig. 4.3.3: Class diagram showing classes which implement the computer email system web application

Class diagrams are the part of Object Oriented analysis and designing, which can be mapped directly with object oriented programming languages like JAVA and C++ etc. Due to this motivation, it is widely used by the software developer community, suggested by James P. Davis et al. in 2007 [21]. At this moment, we implement the code generation and Forward engineering for computer email system in C++ by using IBM Rational Rose. Class wise source code of computer e mail system i.e. component-based programming of e mail case study in C++ is given below. This source code can assist the software analyst, software designers, software developers and software maintenance engineers during the development and maintenance phase of software development life cycle. The class wise equivalent source code in C++ of class diagram (shown in Fig. 4.3.3) of above case study is as follows-

```
Email.h

#ifndef EMAIL_H_HEADER_INCLUDED_ACDE60DF

#define EMAIL_H_HEADER_INCLUDED_ACDE60DF

//##ModelId=532183950221

class Email
```

```
{
   //##ModelId=5321846901AE
   char Subject;
   //##ModelId=5321846B03AC
   char Time;
   //##ModelId=5321846D017B
   Date ComposingDate;
   //##ModelId=5321847300FF
   char Sender;
   //##ModelId=5321847B0053
   char Receiver;
};
#endif /* EMAIL_H_HEADER_INCLUDED_ACDE60DF */
Receiver.h
#ifndef RECEIVER_H_HEADER_INCLUDED_ACDE0320
#define RECEIVER_H_HEADER_INCLUDED_ACDE0320
//##ModelId=5321836F02CB
class Receiver
{
  public:
   //##ModelId=53218452012F
   Login();
   //##ModelId=53218454038F
   Logout();
   //##ModelId=53218457013A
   ReceiveMail();
   //##ModelId=5321845B02F7
   AddContacts();
  private:
   //##ModelId=532184430357
   char Name;
   //##ModelId=53218447017B
```

```
    char ID;
    //##ModelId=5321844801D7
    char Password;
};
#endif /* RECEIVER_H_HEADER_INCLUDED_ACDE0320 */
Receiver.cpp
#include "Receiver.h"
//##ModelId=53218452012F
Receiver::Login ()
{
}
//##ModelId=53218454038F
Receiver::Logout()
{
}
//##ModelId=53218457013A
Receiver::ReceiveMail()
{
}
//##ModelId=5321845B02F7
Receiver::AddContacts(){ }
Sender.h
#ifndef SENDER_H_HEADER_INCLUDED_ACDE29C6
#define SENDER_H_HEADER_INCLUDED_ACDE29C6
//##ModelId=5321836400CA
class Sender
{
  public:
    //##ModelId=532183AD00F8
    Login();
    //##ModelId=532183B00003
    Logout();
```

```
//##ModelId=532183B202EF
    SendMail();
    //##ModelId=5321843800A1
    AddContacts();
  private:
    //##ModelId=532183A00107
    char Name;
    //##ModelId=532183A20159
    char ID;
    //##ModelId=532183A30376
    char Password;
};
#endif /* SENDER_H_HEADER_INCLUDED_ACDE29C6 */
Sender.cpp
#include "Sender.h"
//##ModelId=532183AD00F8
Sender::Login()
{
}
//##ModelId=532183B00003
Sender::Logout()
{
}
//##ModelId=532183B202EF
Sender::SendMail()
{
}
//##ModelId=5321843800A1
Sender::AddContacts()
{
 }
```

Here we have implemented the forward engineering for e mail systems in C++. This source code also we can use for reverse engineering, we verified the object oriented source code in C++. In this roundtrip engineering approach i.e. forward and reverse engineering, the most important advantage is higher flexibility that the developer is not troubled to replicate the process numerous times. As we know, software does not have wear out phase. This source code in C++ can help the software analyst, software designers, software developers and software maintenance engineers during the development and maintenance phase of software development life cycle.

4.4 OBJECT ORIENTED SECURE ROUND TRIP ENGINEERING FOR WEB APPLICATIONS-4

In this article; we are using the experimental software designing and development for Simple answering machine system. UML 2.0 is for the same purpose. No doubt, that software design and development process and object oriented empirical modeling is very important for the development of e-commerce applications and IT applications. Here we considering the Simple Answering Machine, information for the same are given below:—

- Use Case: Leave a Message
- Use Case: Review Messages are the functional requirement, so during analysis, this will be represented as use case in use case diagram
- Actor: The owner will be treated as actor
- Use Case: Review Messages Locally is a functional requirement for the above mentioned application, so during analysis phase, this will be represented as use case in use case diagram
- Use Case: Review Messages Remotely is also functional requirement for the above mentioned application, so during analysis phase, this will be represented as use case in use case diagram
- Use Case: Authorize Access is also functional requirement for the above mentioned application, so during analysis phase, this will be represented as use case in use case diagram.

The projected approach for Simple Answering Machine case study is based on the following phases of the development process:

- First of all, extract web application requirement and requirement discovery through requirement elicitation techniques and Requirement Engineering. This process will help in mining pattern for the system calls.

- Design the use case diagram according to mining pattern of system calls, to represent the functional requirement of web application for the analysis phase of SDLC
- Give details about each use case and write use case description for major use cases
- Specify process behavior using sequence diagrams
- Designing of class diagram and implementation of Forward engineering in C++ source code using IBM Rational Rose for the same secure web application i.e. design phase
- Concept of reverse engineering, re engineering and round trip engineering for the same secure web application.

The Use case diagrams have been drawn to emphasize all important scenarios and cases regarding the system use by these actors. As we know, 'Use Cases' are used to represent the functional requirements and services of the software system. So, we should design the use case diagram of above case study for round trip engineering. In Fig. 4.4.1, we can see that Caller, owner manager etc. are the actor. Rest of the diagram indicates the use cases (for example review caller message, delete caller message, record greeting and set answer mode etc.) i.e. functional requirement of Simple Answering Machine and their relationships. The standard Use case diagram of simple answering machine system is as follows in Fig. 4.4.1:–

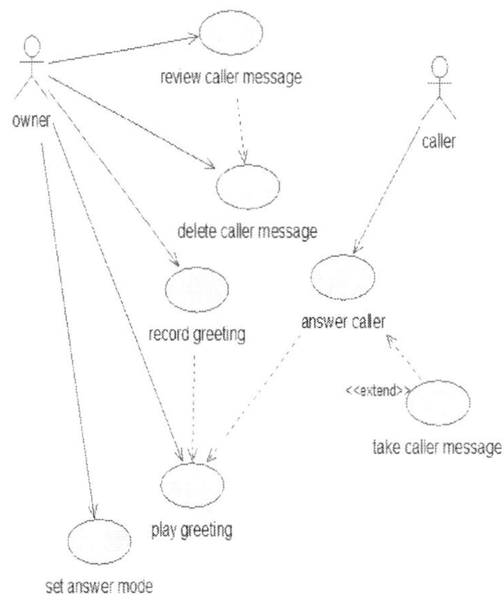

Fig. 4.4.1: Object oriented use case diagram for simple answering machine

According to Fig. 4.4.1, we can write use case description of major Use cases for the understanding of web application. This understanding level can be used during the phases of software development life cycle. These use case description can help in analysis, designing, coding and maintenance etc. Tables 4.4.1 through 4.4.6 show the description of major use cases of simple answering machine.

Table 4.4.1: Use case description of review caller message use case

Brief Description of review caller message use case	**This use case is used to review the message left by caller**
Actors	Owner, Machine
Flow of Events	Owner reviews the messages
Alternative Flow	None
Precondition	The caller has left a message to be reviewed
Post condition	Message is reviewed

Table 4.4.2: Use case description of delete caller message

Brief Description of delete caller message	**This is used to delete the message left by caller**
Actors	Owner, machine
Flow of Events	• Caller selects a message • Caller deletes the message
Alternative Flow	None
Precondition	Message should be present
Post condition	none

Table 4.4.3: Use case description of record greeting use case

Brief Description of record greeting use case	**This use case is user to record a greeting for the caller**
Actors	Owner, Machine
Flow of Events	• Owner press the record button • Owner records a greeting for caller
Alternative Flow	None
Precondition	None
Post condition	The machine has a greeting message for caller

Table 4.4.4: Use case description of play greeting use case

Brief Description of play greeting use case	**This is used to play a greeting when a caller calls and the owner is unable to answer the call**
Actors	Caller, Machine
Flow of Events	• Caller calls the owner • Machine plays the message
Alternative Flow	Caller calls the owner
Precondition	None
Post condition	None

Table 4.4.5: Use case description of set answer mode use case

Brief Description of set answer mode use case	**This use case is used to set the answering mode by owner**
Actors	owner, Machine
Flow of Events	Owner sets the answering mode
Alternative Flow	None
Precondition	None
Post condition	None

Table 4.4.6: Use case description of take caller message

Brief Description of take caller message	**This is used to record the caller's message in the machine**
Actors	Caller, Machine
Flow of Events	Caller calls the owner. An owner provides a message to be recorded. Machine records the message
Alternative Flow	None
Precondition	None
Post condition	None.

For novel modeling and full UML notation, we also can use some stereotypes like <<include>> and <<extend >>. These <<include>> and <<extend >> stereotypes shows the dependencies between use cases. As we can observe in Fig. 4.4.1. This solution can be complemented by designing a sequence diagram. The interaction or sequence diagrams are used when we desire to know the message flow and of the software systems. Currently, message flow means the sequence and ordering of flow from one object to another object. That is why sequence diagram is used to capture dynamic environment but from a special viewpoint. The following are the usages of sequence diagram or interaction diagrams:

- To confine dynamic performance of a software system. As we know, in object oriented language, object will communicate through message passing approach
- To express the communication flow (API, RMI and message flow etc.) in the software system.
- To clarify the structural association of the objects.
- To illustrate interaction between objects.
- No. of Objects taking part in the interaction during interface.
- Message flows (synchronization and a synchronization etc) among the objects in web application.
- Object organization of software systems.

Fig. 4.4.2; demonstrate the sequence diagram for Simple answering machine system.

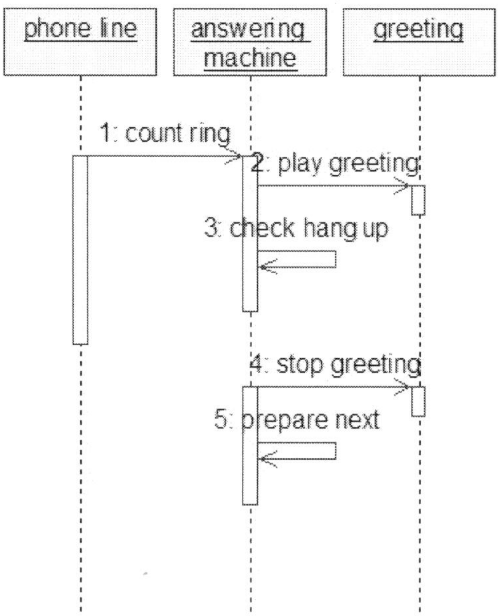

Fig. 4.4.2: Object oriented sequence diagram for simple answering machine

During implementations, Web Developers can implement these Message flows, Object organization, structural association of the objects, ordering of member functions and round trip engineering of software systems. The generic Sequence diagram of projected case study is as pursues in Fig. 4.4.2. Further, Fig. 4.4.3 shows the Object oriented activity diagram for Simple Answering Machine.

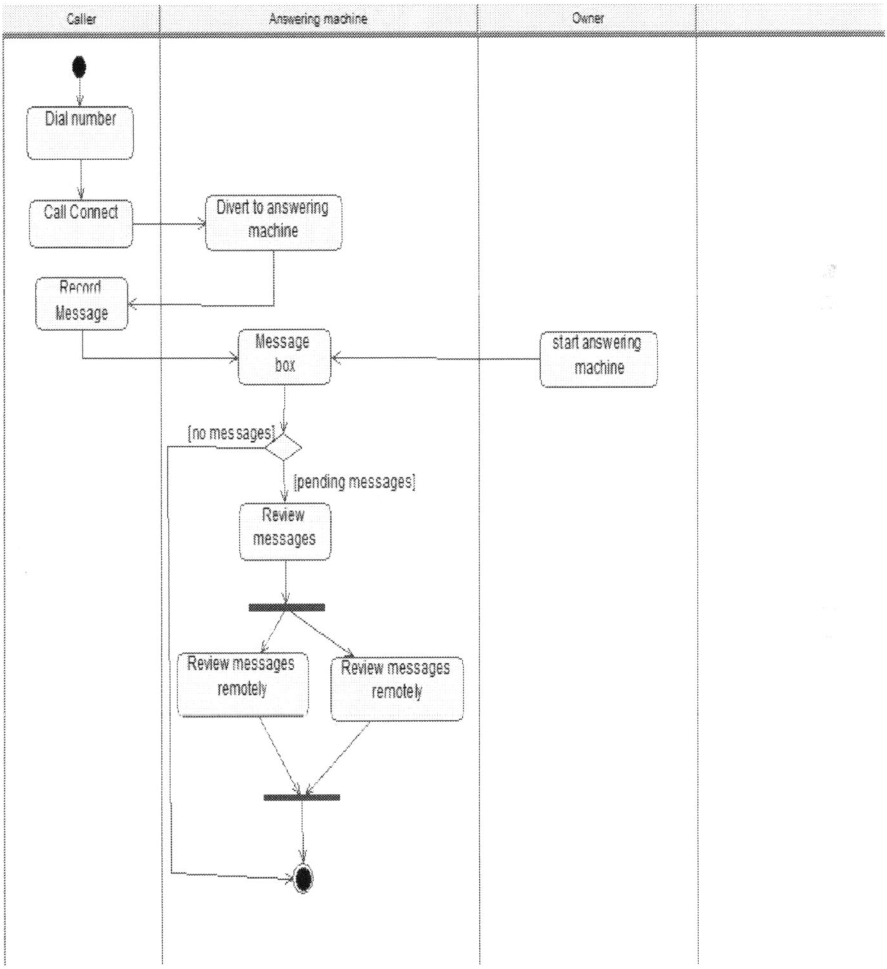

Fig. 4.4.3: Object oriented activity diagram of answering machine

Fig. 4.4.4. Analysis Model of answering machine, which covers all control object, interface object and entity objects of answering machine. In Fig. 4.4.4 calling machine, remote review and local review are interface objects in answering machine. Message recording is the kind of entity object in same case study. Answering controller is the example of control object in answering machine.

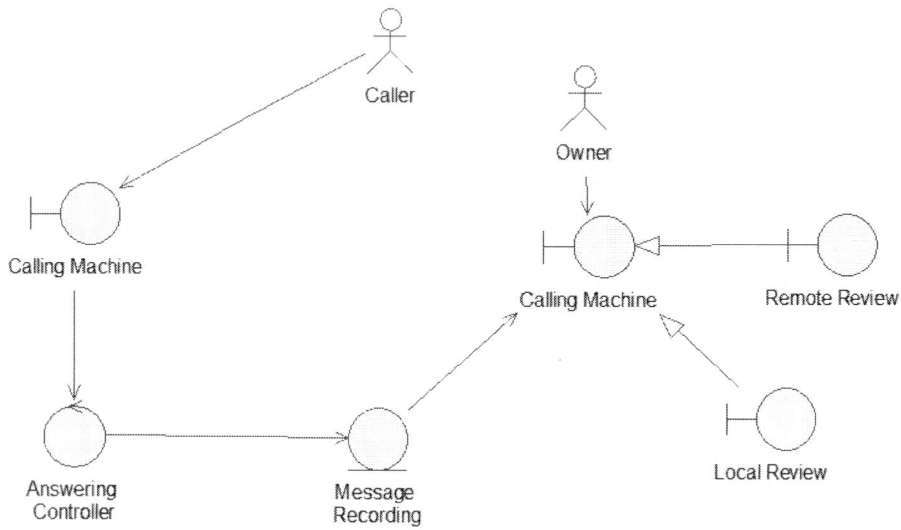

Fig. 4.4.4: Analysis Model of answering machine

This book focuses on empirical relationships between SPF model, the use case, sequence diagram and class diagrams. Web developers require making a well-connected object oriented classes. By using above use case diagram, we expand the component based classes. After this we generate source code in C++ and execute forward engineering and reverse engineering for the same application with the help of forward engineering tool. So from realistic experience class diagram is usually used for construction purpose of software development process. So in a brief, as we know the class diagrams are used for:

- Describing the static analysis of the software system.
- Showing the association and multiplicity between the elements of the static view.

We can gain a better understanding of C++ source code by reading diagram instead of looking to a possibly thousand lines of source code. The standard class diagram of Simple Answering Machine application is as follows in Fig. 4.4.5.

Logical and technical Designers can build a high level object oriented class model, then pass to web developer to perform more lower-level software system modeling, and ultimately follow the forward engineering to generate source code. With IBM rational Rose, we can simply generate source code in object oriented language from class diagram. This source code will work as blue print for software development team.

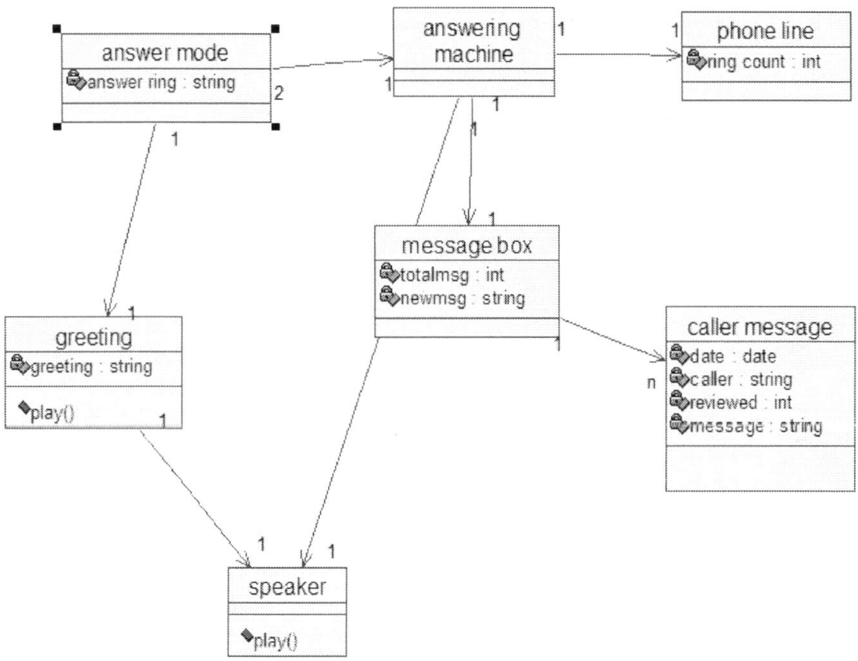

Fig. 4.4.5: Object oriented class diagram for online simple answering machine

Here we recommend the implementation of forward engineering in C++ for simple answering machine based application. In this Simple Answering Machine Based Application, storing objects may be sales answering machine, caller, receiver, owner, phone line and message box etc. (See Fig. 4.4.5). The class wise equivalent C++ Source Code for round trip engineering is as follows:-

```
#ifndef ANSWER_MODE_H_HEADER_INCLUDED_B0857F86
#define ANSWER_MODE_H_HEADER_INCLUDED_B0857F86
//##ModelId=4F7AE1B70213
class answer mode
{
 //##ModelId=4F7AE1C2009C
   string answer ring;
};
#endif
 /* ANSWER_MODE_H_HEADER_INCLUDED_B0857F86 */
#ifndef ANSWERING_MACHINE_H_HEADER_INCLUDED_B0854EA4
```

```
#define ANSWERING_MACHINE_H_HEADER_INCLUDED_B0854EA4
//##ModelId=4F7AE1CB000F
class answering machine
{
};
#endif /* ANSWERING_MACHINE_H_HEADER_INCLUDED_B0854EA4 */
#ifndef CALLER_MESSAGE_H_HEADER_INCLUDED_B0854F00
#define CALLER_MESSAGE_H_HEADER_INCLUDED_B0854F00
//##ModelId=4F7AE20D00FA
class caller message
{
   //##ModelId=4F7AE302032C
   date date;
   //##ModelId=4F7AE30501D4
   string caller;
   //##ModelId=4F7AE3080222
   int reviewed;
   //##ModelId=4F7AE30F01F4
   string message;
};
#endif
/* CALLER_MESSAGE_H_HEADER_INCLUDED_B0854F00 */
#include "greeting.h"
//##ModelId=4F7AE1F7006D
greeting::play()
{
}
#ifndef GREETING_H_HEADER_INCLUDED_B085664F
#define GREETING_H_HEADER_INCLUDED_B085664F
//##ModelId=4F7AE1E90232
class greeting
```

```
{
  public:
    //##ModelId=4F7AE1F7006D
    play();
  private:
    //##ModelId=4F7AE1EE01E4
    string greeting;
};
#endif
/* GREETING_H_HEADER_INCLUDED_B085664F */

#ifndef MESSAGE_BOX_H_HEADER_INCLUDED_B0851094
#define MESSAGE_BOX_H_HEADER_INCLUDED_B0851094
//##ModelId=4F7AE1D2008C
class message box
{
   //##ModelId=4F7AE1DD0251
   int totalmsg;
   //##ModelId=4F7AE1E103A9
   string newmsg;
};
#endif /* MESSAGE_BOX_H_HEADER_INCLUDED_B0851094 */
#include "speaker.h"
//##ModelId=4F7AE31F006D
speaker::play()
{
}
#ifndef SPEAKER_H_HEADER_INCLUDED_B0853B19
#define SPEAKER_H_HEADER_INCLUDED_B0853B19
//##ModelId=4F7AE318031C
class speaker
{
 public:
```

```
    //##ModelId=4F7AE31F006D
    play();
};
#endif /* SPEAKER_H_HEADER_INCLUDED_B0853B19 */
#ifndef PHONE_LINE_H_HEADER_INCLUDED_B0855727
#define PHONE_LINE_H_HEADER_INCLUDED_B0855727
//##ModelId=4F7AE1FD00CB
class phone line
{
    //##ModelId=4F7AE2030128
    int ring count;
};
#endif /* PHONE_LINE_H_HEADER_INCLUDED_B0855727 */
```

By roundtrip engineering, we have developed the understanding level of complex software. In the above experimental case, we have transformed the object oriented models into C++ source code. With the help of above software development process, developers can identify software Metrics like no. of data member per class, friend functions and friend classes, data structure metrics, coupling between objects, information flow and many more.We can also calculate the vocabulary of a program, the length of the program, Volume, potential Volume, Program Level, Program Difficulty, Effort, Count of executable statements, average number of live variables and the total number of variables etc.This forward engineering will also help the developer to know about the degree of cohesion and coupling.

Here we suggest the transition process of object oriented forward and reverse engineering. As we all know that Round-trip engineering is the capability of creating object oriented UML models from the source code and generating source code from these object oriented UML model, and keeping them synchronized. For any web application, we can use the round-trip engineering to make our implementation model and source code up-to-date. Object oriented design information and some obtainable design documentation are improved from the Source Code. The procedure of improving a program's design is known as design recovery. A design is improved by piecing information equally from the source code, obtain documents, and software developer's experience is enhanced with the software system knowledge.

As we know, the software roundtrip engineering i.e. forward engineering and backward engineering (as we can notice in Fig. 4.4.6 and Fig. 4.4.7) plays a vital role in software development life cycle. In this book, we have used round trip engineering implementation for the same simple answering machine system. As we know, in Reverse Engineering, we study the complete system to recognize all the components of the system and their interrelationships so as to ascertain which components can be reused, and how they can be reused.

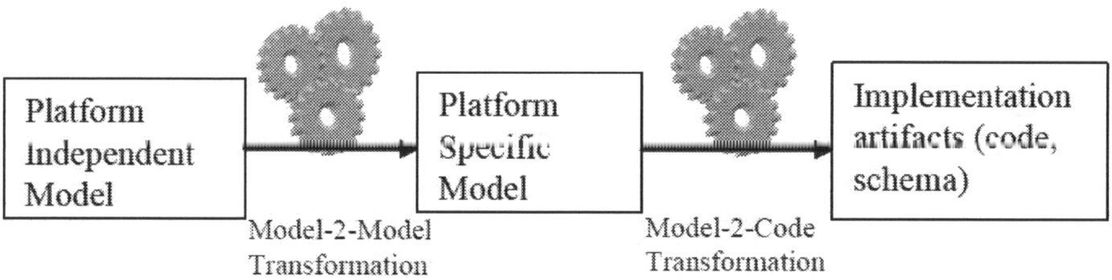

Fig. 4.4.6: Transformation approach in forward engineering.

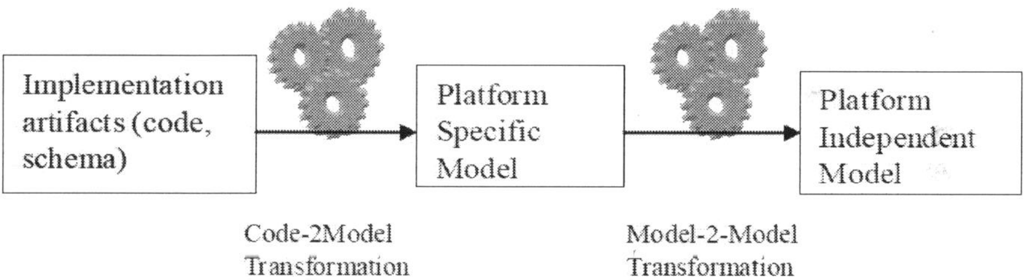

Fig. 4.4.7: Transformation approach in reverse engineering.

Several business object oriented tools are there which provide the reverse engineering abilities. In this book we have used IBM Rational Rose for the same purpose. The information can further be used to develop and preserve other software systems; i.e. we can acquire supplementary information that can be used by a forward engineer for the purpose of forward engineering. UML tools mostly apply reverse engineering to execute round-trip engineering. They convert source code back into a model. We can implement this suggested approach for any web application.

Let's take an example that is related to the automation of medical sector. In this example, a Patient can place as well as cancel appointment with doctor or physician using scheduler. Doctor or Physician decides to recommend Medication for Patient. Physician Specifies Drug Info:

Medication Name, Dosage Amount and number of Doses etc. Computer Cross-Checks for inconsistency among Medication and recent Medical History Prescription Forwarded Electronically to Pharmacy or Else Printed for Patient. The complete code/ blue print of this application and the proposed approach of this example is provided in Appendix I. Appendix I is covering the mining and SPF based round trip engineering for the automation of medical industry. In appendix I, we can see several use cases in Fig. I-A i.e. the functional requirements or services for above real world case study. Use cases like Arrange Appointment, Cancel Appointment, Medical History, Print Medication, cross check for conflicts and it forwards it to the pharmacy etc. These services or functional requirements are identified during the requirement elicitation and analysis phase of software development life cycle.

Table I-A, Table I-B, Table I-C, Table I-D, Table I-E, Table I-F and Table I-G shows the use case description of major use cases used in the above case study. In appendix I, Fig. I-B shows the analysis model with interface object, control object and entity object. At the time of design phase, we can identify various object oriented classes like Patient, Physician, Computer and scheduler. Every class will have possible data members and its member functions, as we can see in Fig. I-C. In appendix I, we are demonstrating the forward engineering or code generation in C++ language for above mentioned real world case study.

5

Performance Evaluation of Security in Applications

This Chapter enlists the results of several measures used to estimate the possible performance enhancement due to SPF application. Due to good performance of SPF, we are suggesting the same for the implementation of web applications. The selection of SPF model for applications will depend upon the web application requirement. In some application, system based SPF can be used. In other applications other SPF model may be beneficial.

For example, Web server deals with public information. Since majority of data and information is freely available to all users on web server, task of checking it during read from disk is something useless because this data is already readable by each and every user using internet. So in this example, admin of the system can disable all the read checks in web server because they are actually useless which finally increases throughput of the web server. The real task of security comes when it comes to writing access. This means, in above application, web server integrity is the main issue rather than its confidentiality. By skipping these processes or system calls the performance of system will be improved. These SPF based improved version of secure systems can be used for desired web application. This selection of operations and system calls will be different from one application to another. So for the selection of these necessary and un-necessary operations pattern mining is required. We have done this pattern mining and development of system for some real life applications in chapter 4.

We propose SPF model to measure the throughput of different system calls and I/O operations, UnixBench version 4.0.1 microbenchmark has been applied. The different benchmark parameters applied for performance efficiency evaluation were executed on Redhat, SELinux without SPF, SELinux with system based SPF and SELinux with process based SPF kernels. The results will analyze to predict how efficiency SPF implementation was able to enhance the performance of SELinux when compared to Redhat. In all web application, all security checks are not essential. We can skip or disable some un-necessary security check from web application. With this approach, the performance of web applications will be improved in all respect. Implementation of SPF is needed for high performance in web application and this is possible through forward engineering.

After collection and analysis of benchmark results, we observed the performance improvement of SPF model. After analysis, we also identified the performance differences in the System based SPF and Process based SPF implementations. Although, both the SPF framework implemented in a different way. The implementation of system and process SPF will vary from one web application to another. Hence, it is complex to estimate if any particular implementation has a performance benefit over the other. The table below demonstrate the results of SPF based System and Process based SPF. These results are based on disabling of security checks. If we skip some un-necessary security checks on operation than the performance of SPF well be improved. It is observed that there is small performance variation between the above mentioned system-SPF and process-SPF implementations.

Table 5.1 to Table 5.6 demonstrates the results from Unix Bench microbench marks significant to SPF. Table 5.1 and Table 5.4 explain the outcomes of the AIM9 and UnixBench with the executed security. Table 5.2 and table 5.5 demonstrate the outcomes of the microbenchmarks with the skipped security checks. Table 5.2 and table 5.5, the appropriate SPF design is set so the microbenchmark operation skips the security checks. Table 5.3 and table 5.6 enlist the performance degradation percentage of SELinux without SPF. The result table also highlights the performance enhancement that each SPF implemented kernel is capable to keep from the original SELinux performance loss.

The AIM9 benchmarks have been removed from the system SPF and Process based SPF columns of Table 5.1 to table 5.3. These all results are based on the AIM9 benchmarks. The numbers calculated are then averaged to results in Table 5.1, Table 5.2, Table 5.4, and Table 5.5. The results obtained prove that SELinux security can have an important performance effect on the operations studied. Table 5.3 and Table 5.6 show the performance degradation percentage of SELinux without SPF and Redhat. The percentage loss results are reasonably important. Performance loss has been noted to be as high as 41.89% in one case, but most of the tests had performance loss ranging from 5% to 15% roughly, which is comparatively important and noticeable.

Data in Table 5.1 to Table 5.6 specify that the overhead of SPF mechanism is relatively irrelevant. Table 5.3 and Table 5.6 demonstrate that all the benchmark tests, SPF are successfully regaining all the performance loss which occurred due to SELinux security. The percentage improvement numbers as shown in Table 5.3 and Table 5.6 signify that SPF overhead lowers the performance of these microbenchmarks by roughly 1% to 2%. In general, the results are as hypothesized. The SPF overhead is negligible an insignificant as compared to SELinux security. This significant overhead variation enables the attainment of high performance through the application of SPF frame work.

Table 5.1: AIM9 Benchmark Results, security checks executed.

File System Tests	Redhat	SELinux No SPF	SELinux System-SPF	SELinux Process-SPF
Random Disk Reads (K)/second	100547.12	94167.04	93135.59	94531.9
Random Disk Writes (K)/second	84529.51	79188.24	79508.83	80496.64
Sequential Disk Reads (K)/second	371456.00	335527.37	325591.04	332840.96
Sequential Disk Writes (K)/second	159010.60	149616.64	153174.03	152419.35
Disk Copies (K)/second	110075.60	102252.55	102744.05	103895.04
Directory Searches/second	65973.45	38335.35	37657.05	
Directory Operations/second	2756600.00	2705540.00	2698520.00	

Table 5.2: AIM9 Benchmark results, security checks skipped.

File System Tests	Redhat	SELinux No SPF	SELinux System-SPF	SELinux Process-SPF
Random Disk Reads (K)/second	100547.12	94167.04	99762.33	100403.2
Random Disk Writes (K)/second	84529.51	79188.24	84768.48	85123.42
Sequential Disk Reads (K)/second	371456.00	335527.37	363571.20	361014.22
Sequential Disk Writes (K)/second	159010.60	149616.64	159727.37	160055.67
Disk Copies (K)/second	110075.60	102252.55	110315.52	109041.4
Directory Searches/second	65973.45	38335.35	40259.85	-
Directory Operations/second	2756600.00	2705540.00	2753680.00	-

Table 5.3: AIM9 Benchmark improvement, security checks skipped.

File System Tests	SELinux No SPF Performance Degradation	SELinux System-SPF Improvement Over SELinux No SPF	SELinux Process-SPF Improvement Over SELinux No SPF
Random Disk Reads (K)/second	-6.35%	+5.57%	+6.21%
Random Disk Writes (K)/second	-6.32%	+6.60%	+7.02%
Sequential Disk Reads (K)/second	-9.67%	+7.55%	+6.86%
Sequential Disk Writes (K)/second	-5.91%	+6.36%	+6.57%
Disk Copies (K)/second	-7.11%	+7.33%	+6.17%
Directory Searches/second	-41.89%	+2.91%	-
Directory Operations/second	-1.85%	+1.74%	-

Table 5.4: UnixBench Benchmark results, security checks executed

File System Tests	Redhat (KB/sec)	SELinux No SPF (KB/sec)	SELinux System-SPF (KB/sec)	SELinux Process-SPF (KB/sec)
File Read 256B Buffer	164178.20	137348.80	136456.20	137537.8
File Write 256B Buffer	160825.00	135722.80	130492.80	133475.6
File Copy 256B Buffer	76769.60	64952.60	63179.20	64980.8
File Read 1KB Buffer	367885.20	335115.00	328872.40	331970.8
File Write 1KB Buffer	200432.40	189583.60	191452.80	191578.8
File Copy 1KB Buffer	129604.60	120578.20	120877.80	121285.8
File Read 4KB Buffer	538517.40	516731.80	515341.60	514826.8
File Write 4KB Buffer	224121.60	221207.60	224421.20	224154.8
File Copy 4KB Buffer	156394.40	152110.20	154932.60	154295.2

Table 5.5: UnixBench Benchmark results, security checks skipped.

File System Tests	Redhat (KB/sec)	SELinux No SPF (KB/sec)	SELinux System-SPF (KB/sec)	SELinux Process-SPF (KB/sec)
File Read 256B Buffer	164178.20	137348.80	161259.60	159056.4
File Write 256B Buffer	160825.00	135722.80	158814.60	155229.2
File Copy 256B Buffer	76769.60	64952.60	75257.40	75209.0
File Read 1KB Buffer	367885.20	335115.00	365526.60	364978.8
File Write 1KB Buffer	200432.40	189583.60	204503.80	205676.8
File Copy 1KB Buffer	129604.60	120578.20	130455.60	130986.4
File Read 4KB Buffer	538517.40	516731.80	536233.60	533637.2
File Write 4KB Buffer	224121.60	221207.60	229098.60	231857.0
File Copy 4KB Buffer	156394.40	152110.20	158921.00	158012.8

Table 5.6: UnixBench Benchmark improvement, security checks skipped.

File System Tests	SELinux No SPF Performance Degradation	SELinux System-SPF Improvement Over SELinux No SPF	SELinux Process-SPF Improvement Over SELinux No SPF
File Read 256B Buffer	-16.34%	+14.56%	+13.22%
File Write 256B Buffer	-15.61%	+14.36%	+12.13%
File Copy 256B Buffer	-15.39%	+13.42%	+13.36%
File Read 1KB Buffer	-8.91%	+8.27%	+8.12%
File Write 1KB Buffer	-5.41%	+7.44%	+8.03%
File Copy 1KB Buffer	-6.96%	+7.62%	+8.03%
File Read 4KB Buffer	-4.05%	+3.63%	+3.14%
File Write 4KB Buffer	-1.30%	+3.52%	+4.75%
File Copy 4KB Buffer	-2.74%	+4.36%	+3.77%

The tabulated results give no clear sign or indication of better performance between any SPF implementation (System-SPF and Process-SPF) however; the AIM9 benchmark indicates that System based SPF or Process based SPF has lower overhead than File-SPF. Hence, we can conclude that the performance overhead of all SPF implementations under study is more or less equal. To maintain the security in web applications, we can use all the SPF implementation. In this book, we have described experiential round trip engineering, object oriented modularization, source code structuring and restructuring of secure software system.

As we know, before implementation modeling is required to design and develop the complex software. Implementation of SPF model is a part of system programming and system programming is so complex. Therefore the modeling of SPF is essential before the real implementation and development. Here in this work, for modeling purpose we propose object oriented forward and reverse engineering through IBM rational Rose software. The selection of above mentioned SPF will vary from one web application to another one. The main contribution of this work is that not the entire security in Trusted Operating System is required. Further skipping irrelevant security checks in such System may improve the performance of system or web application in all respect. This SPF configuration for applications provides administrators the power to balance security and performance needs of a system as per the requirement of users. The SPF model can be designed and implemented at various levels like system based SPF and process based SPF implementation.

This work creates an experiential base to showcase the effectiveness and efficiency of SPF based object oriented modeling mechanism in software development, with primary focus on security in applications. To retain the balance among security and web application performance, we propose amalgamation or combination of security based forward engineering and round-trip engineering. This round-trip engineering in software development is used to maintain synchronization between source code and software design. This work proposes SPF based secure software analysis, SPF based secure software design, SPF based secure software development and roundtrip engineering based maintenance approach. This book also demonstrates how real-life product-line products can be built on top of this platform. We have considered some real life case studies for web applications in chapter 4 for the same proposed scheme. To maintain the legacy of source code, this work also proposes that round trip engineering is necessary for real life applications. Therefore, one should not only focus on the program development but also on system maintenance. Because software and web application do not have wear out phase. Application development is short term process but maintenance, reverse engineering, reengineering and restructuring of source code is long term process and much time consuming. Source code listing and Restructuring of program should be done in such a way that it completely avoids the overhead of reverse engineering of a system again and again because of modifications being made to its code over time.

6

Conclusions and Future Work

6.1 CONCLUSIONS

This book is nothing but result of an extensive investigative study on this work and one small step in the ongoing journey of an endless exploration. As we know, before implementation modeling is required to design and develop the complex software. Implementation of SPF model is a part of system programming and system programming is so complex. Therefore the modeling of SPF is essential before the real implementation and development. Here in this work, for modeling purpose we propose object oriented forward and reverse engineering through IBM rational Rose software. This book explains UML-based foundations for model driven architecture and forward engineering of UML models. In this projected process, we design object oriented models like use case diagram, sequence diagram, class diagram, etc. We use object-oriented conceptual modeling techniques to design and develop various applications like social networking sites, E-commerce, Comparison shopping and product purchase system, etc.

This book suggests forward engineering to generate source code from object models through IBM Rational Rose software. This object oriented source code help software development team from analysis to maintenance phase as well as for round trip engineering. Due to high security concern, we have to use highly secure operating systems as a platform to run these web applications. The novel integration of security engineering with model driven Software expansion approach is being used in this book.

This work proposes a reverse engineering mechanism for source code restructuring, reconstructing the design of an implemented software system. This approach was earlier applied to reconstruct and document the design of a software program. However, it is observed that the changes in the software design easier than the source code. The design recovered through reverse engineering explains the current software source code and software system. The resultant high-level design will be updated to generate a new detailed design. A new version or reengineering of the system is obtained from this design. The new version of software can be developed through varied recent techniques and programming languages, resulting is a better structured documented as well as maintainable system. Hence, reverse engineering as a part of re-engineering will prove to be more useful in usability and life of legacy code. During the software development, for maintaining the security concern, SPF model is suggested.

The main contributions of this work are given below:

- The original proposal that not the entire security in Trusted Operating System is required.
- The fresh initiative that skipping irrelevant security checks in such System may improve the performance of system in all respect.
- This SPF structure of TOS provides administrators the privilege to balance security and performance needs of a system as per the requirement of users.
- The SPF model can be designed and implemented at various levels like system based and process based implementation is possible. This selection will be under consideration according to system's need.
- This work allows system administrators to configure and arrange such systems dynamically as quickly as the system executes.
- This work proposes SPF based secure software analysis, SPF based secure software design, SPF based secure software development and roundtrip engineering based maintenance approach. We have also demonstrated how real-life product-line products can be built on top of this platform. We have already taken some case studies in chapter 4 for the same proposed scheme. It is necessary to perform reverse engineering to maintain the legacy of source code. Therefore, one should not only focus on the program development but also on system maintenance. Because software and web application do not have wear out phase. Application development is short term

process but maintenance, reverse engineering, reengineering and restructuring of source code is long term process and much time consuming. Source code listing and Restructuring of program should be done in such a way that it completely avoids the overhead of reverse engineering of a system again and again because of modifications being made to its code over time.

- This work creates an experiential base to showcase the effectiveness and efficiency of SPF based object oriented modeling mechanism in software development, with primary focus on security in applications. To retain the balance among security and web application performance, we propose amalgamation or combinbnation of security based forward engineering and round-trip engineering. This round-trip engineering in software development is used to maintain synchronization between source code and software design.
- The analysis and data mining of huge repositories includes chances of unwanted disclosure to the sensitive data or private information which pertains to a particular business organization. In the recent years, the importance to prevent unwanted disclosure of data patterns or information by an unauthorized means have been increased tremendously and have become most demanding too. The mining restrictions imposed by proposed information securing system, includes complex processing steps, so it becomes a necessity to visualize and design its model accordingly before the actual implementation is done.

In this book, we have described experiential round trip engineering, object oriented modularization, source code structuring and restructuring of secure software system. We have analyzed the quality of the system in all and have gained an outline of it in terms of its size, structure and complexity. Then we locate and understand the most important classes and hierarchies and Message Passing Coupling in classes etc. This book illustrates the abstract cohesion of classes through forward engineering in C++. In this work, we have presented a SPF based software development process for web applications. We have also analyzed the integration of security engineering into a model-driven software development approach. Object-oriented and SPF based conceptual modeling techniques applied to the development of online web applications.

6.2 FUTURE WORK

This book being a work has been constrained with the time like any other such tasks. Hence, the activity of investigation appears endless. Some of the tasks are given here that we intend to take up in future for work investigation.

Still there exists many open issues with respect to the approach; they are both theoretical as well as practical. The study of SPF model can be explored more in near future. The SPF design can be implemented for some more real life web applications. In this work, we explored the SPF design at system and process level. The implementation of SPF can be possible at file level. The file based SPF can be considered for future work. As it is revealed that in near future SPF can be implemented for file system of SELinux. This can be taken further and can be done with networking or even with IPC. Implementation of SPF within the networking subsystem of a Trusted Operating System may help decrease the latency of network messages.

Mostly, this work should be considered an evaluation of all the possibilities for SPF in Trusted Operating Systems. Still much more is awaited for further improvement. At the time of writing this work, Trusted Operating Systems are still not commonly used for web applications. Future in near future these operating systems can be used broadly. Further implementation of SPF is possible for some other real life e-commerce applications. In this work, we designed process and system SPF for some case studies. But more depth knowledge can be explored in near future.

For analysis and designing of SPF for e commerce applications, we have proposed SPF based round trip engineering for web application. Currently our design is focusing on static i.e. the fixed design models, which are relatively very close to the object oriented implementation. Our Future work can focus on modeling security requirements and design information using the dynamic UML models. We also need to obtain many more experimental evidences than we have now and some more evidences about the applicability of the integration strategies on more real-world examples.

7

Solved Case Studies

Q.1) Consider a software system for supporting a Public Library

- Identify actors for Library System. Explain the relevance of each actor.
- One use case is to borrow a library item. List three additional use cases at a comparable level of abstraction. Describe each use case with the exceptional flow.
- Prepare a Use Case Diagram (UCD) for a library system. Describe major use cases with exceptional flow.
- Elaborate Library Item as an abstract class with concrete classes class.
- Identify at least four states of Library item and draw the State Transition Diagram for the same.

Answer 1:

The various participants of the same are detailed below: -

Actors: - Member, Librarian

The corresponding use cases for these actors are:-

- **Member**: Inquiry for Membership, Search Book, Book Issue, Book Return, Pay Fine
- **Librarian**: Search Book, Issue Membership Card, Cancel Membership, Issue Book, Return Book, Charge Fine in Case of Late Return, Maintain the Book Records, Add Books, Remove Books, Add Members, Remove Members, Update Member Details

Here we have some dependencies also like Request for Book Return <<**extends**>> Pay Fine. Also Maintain Book Records <<**includes**>> Add

Book, Remove Book and Update Book. Again Return Book <<**extends**>> Charge Fine in case of late return.

The possible UML Use Case diagram for Library Management System is shown below: -

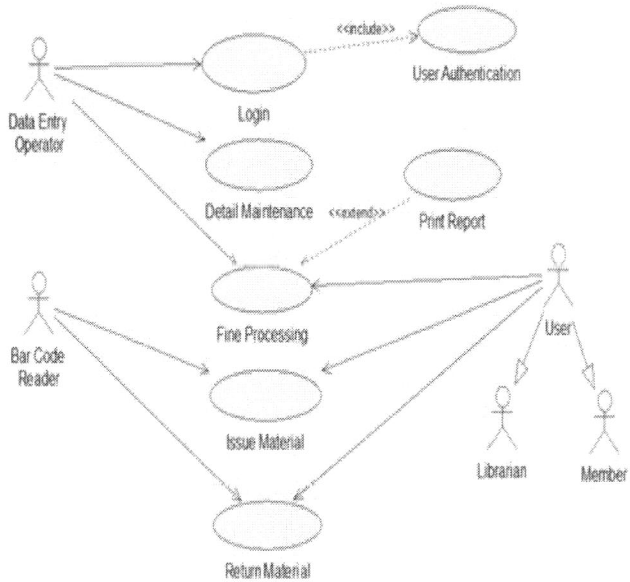

Use Case Descriptions of Major Use Cases

Login Use Case

Brief Description	This use case describes how a user login to the Public Library System
Actors	User (of generalized types passenger and agent) , data entry operator, bar code reader
Flow of Events	1. Enter user id 2. Enter password 3. Click on login button
Alternative Flow	If he isn't an existing user then an option to sign up.
Precondition	Passenger/agent must have an valid account.
Post condition	If the use case was successful the actor is logged in to the system after receiving a success message. If not, the system state is unchanged.

User Authentication Use Case

Brief Description	This use case verifies if the user is an authentic user of the Public Library System.
Actors	User (of generalized types passenger/agent) , data entry operator, bar code reader
Flow of Events	1. Enter user id 2. Enter password 3. Click on login button
Alternative Flow	If he isn't an existing user then an option to sign up.
Precondition	passenger/agent must have entered his login credentials
Post condition	If the use case was successful the actor is logged in to the system after receiving a success message. If not, the system state is unchanged.

Detail Maintenance Use Case

Brief Description	This use case enables the data entry operator to maintain details of the library in the Public Library System.
Actors	Data entry operator
Flow of Events	1. Log into the system 2. Perform maintenance task
Alternative Flow	If the public library system isn't working, intimate the user.
Precondition	The data entry operator must have already logged in.
Post condition	If the use case was successful the actor is able to perform maintenance operation. If not, the system state is unchanged.

Fine Processing Use Case

Brief Description	This use case enables the data entry operator to process member fine in the Public Library System.
Actors	Data entry operator
Flow of Events	1. Log into the system 2. Check for defaulters 3. Calculate fine
Alternative Flow	If the public library system isn't working, intimate the user.
Precondition	The data entry operator must have already logged in.
Post condition	If the use case was successful the actor is able to perform fine processing operation. If not, the system state is unchanged.

Issue material Use Case

Brief Description	This use case enables the bar code reader to issue material in the Public Library System.
Actors	Bar code reader
Flow of Events	1. Authenticated by system 2. Read barcode 3. Issue material
Alternative Flow	If the public library system isn't working, intimate the user.
Precondition	The bar code reader must be authenticated by the system.
Post condition	If the use case was successful the actor is able to issue material from the system. If not, the system state is unchanged.

The possible Class Diagram for Library Management System are shown below: -

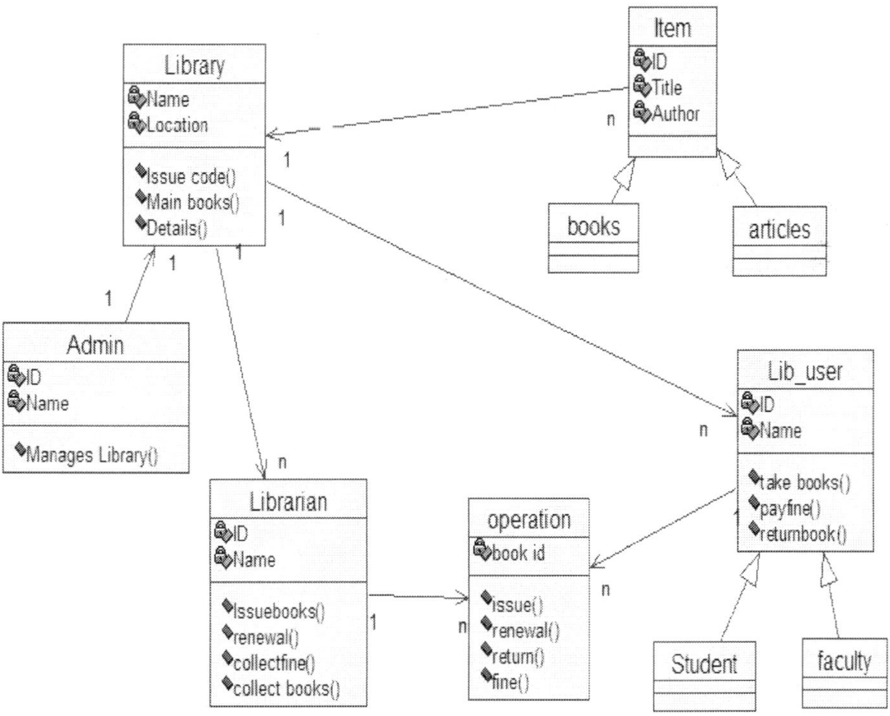

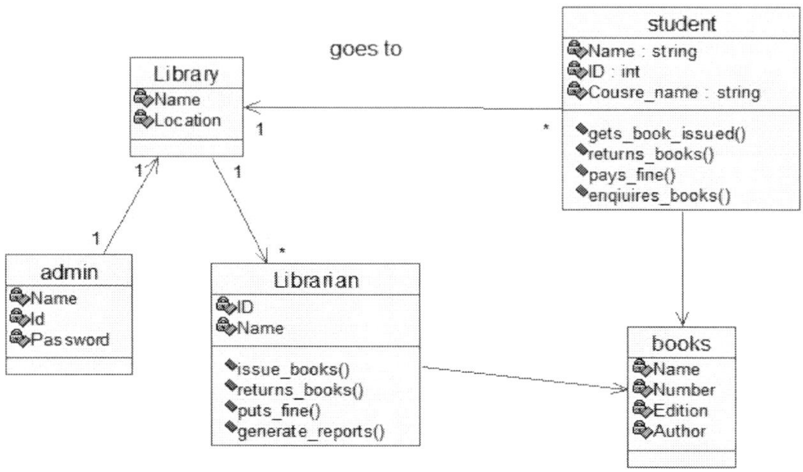

The State Diagram for Library Management System is shown below:-

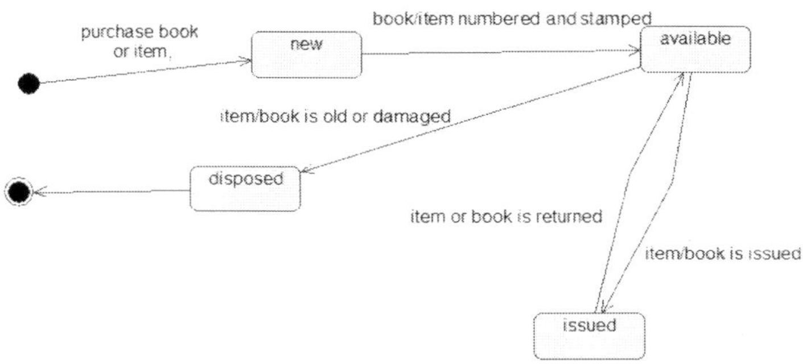

Q. 2) Consider an Online Airline Reservation System. You may want to check airline web sites to give you idea.

- Identify actors for Online Airline Reservation System. Explain the relevance of each actor.
- One use case is to make a Flight Reservation. List four additional use cases at a comparable level of abstraction. Describe each use case with exceptional flow.
- Prepare a **Use case Diagram** for an online airline reservation system. Write **Use case description of each use case.**
- Draw **Class Diagram**. Elaborate **ticket class and Generate Code** for the same.
- Identify at least four states of Ticket object and draw the **State Transition Diagram** for the same.

Answer 2:

IDENTIFICATION OF USE CASES AND ACTORS:

ACTORS:

1. Reservation Clerk.
2. Passenger.

USE CASES:

1. Maintaining Flight Details.
2. Maintaining Flight Status.
3. Reserve Tickets
4. Cancellation Tickets

Possible Use case Diagram of Online Airline Reservation System:

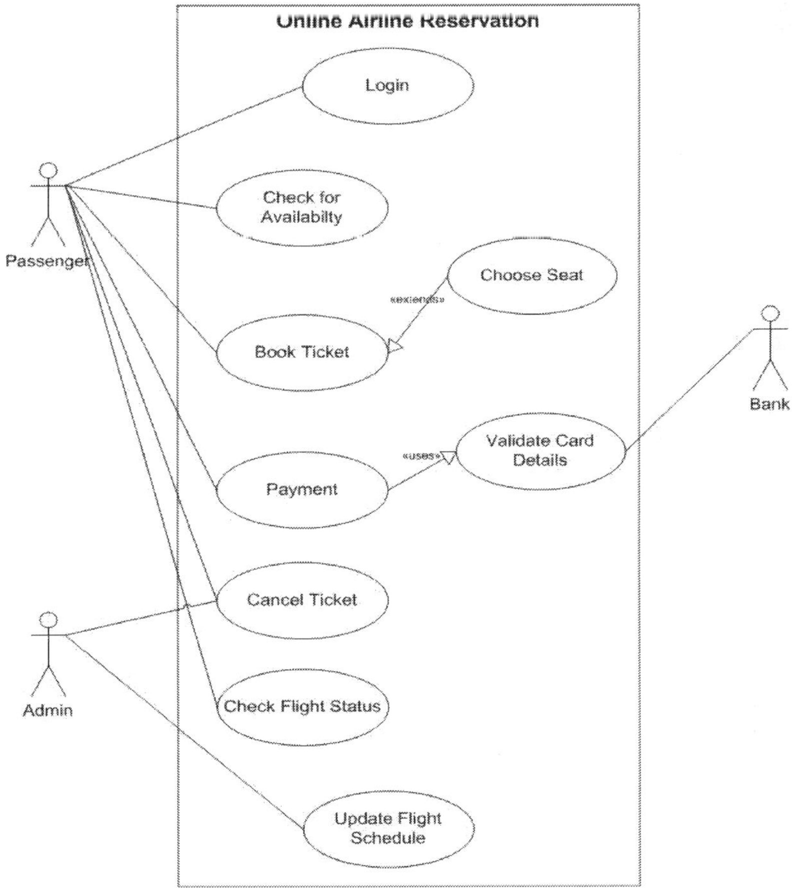

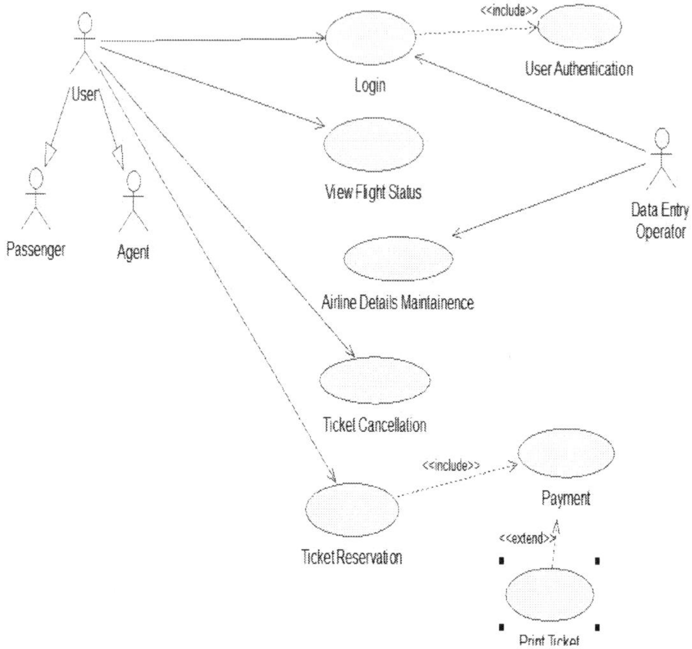

Use Case Descriptions of Major Use Cases

Login Use Case

Brief Description	This use case describes how a user login to the Airline Reservation System
Actors	User (of generalized types passenger and agent) , data entry operator
Flow of Events	1. Enter user id 2. Enter password 3. Click on login button
Alternative Flow	If he isn't an existing user then an option to sign up.
Precondition	Passenger/agent must have an valid account.
Post condition	If the use case was successful the actor is logged in to the system after receiving a success message. If not, the system state is unchanged.

User Authentication Use Case

Brief Description	This use case verifies if the user is an authentic user of the Airline Reservation System.
Actors	User (of generalized types passenger/agent) , data entry operator

Flow of Events	1. Enter user id 2. Enter password 3. Click on login button
Alternative Flow	If he isn't an existing user then an option to sign up.
Precondition	passenger/agent must have entered his login credentials
Post condition	If the use case was successful the actor is logged in to the system after receiving a success message. If not, the system state is unchanged.

View Flight status Use Case

Brief Description	This use case describes how a user to view flight status the Airline Reservation System
Actors	User (of generalized types passenger and agent)
Flow of Events	1. Log in to the system 2. Select flight details 3. View flight status
Alternative Flow	If the site is down, user shall be intimated.
Precondition	Passenger/agent must have an valid account.
Post condition	If the use case was successful the actor is able to view a flight status. If not, the system state is unchanged.

Airline Detail Maintenance Use Case

Brief Description	This use case enables a data entry operator to maintain the Airline Reservation System
Actors	Data entry operator
Flow of Events	1. Log in to the system 2. Perform maintenance operation
Alternative Flow	If the site is down, user shall be intimated.
Precondition	The data entry operator must have an valid account.
Post condition	If the use case was successful the actor is able to successfully perform maintenance operation. If not, the system state is unchanged.

Ticket Cancellation Use Case

Brief Description	This use case enables a user to cancel a ticket in the Airline Reservation System
Actors	User (of generalized types passenger and agent)
Flow of Events	1. Log in to the system 2. Enter registration id 3. Cancel ticket

Alternative Flow	If the site is down, user shall be intimated.
Precondition	Passenger/agent must have already booked a ticket.
Post condition	If the use case was successful the actor is able to successfully cancel his/her booking. If not, the system state is unchanged.

Ticket Reservation Use Case

Brief Description	This use case enables a user to book a ticket in the Airline Reservation System
Actors	User (of generalized types passenger and agent)
Flow of Events	1. Log in to the system 2. Enter journey details 3. Enter basic details 4. Book a ticket
Alternative Flow	If the site is down, user shall be intimated.
Precondition	Passenger/agent must have already logged into the system.
Post condition	If the use case was successful the actor is able to successfully book a ticket. If not, the system state is unchanged.

Class Diagram of Airline Reservation System

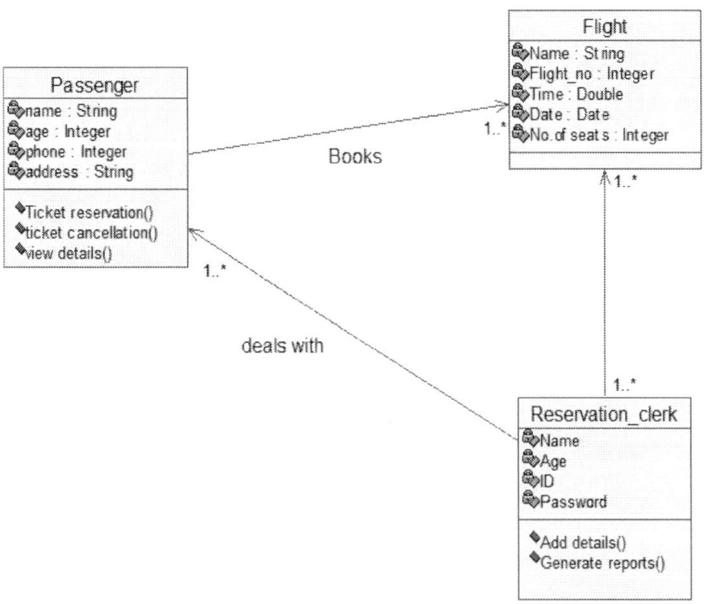

State Diagram of Airline Reservation

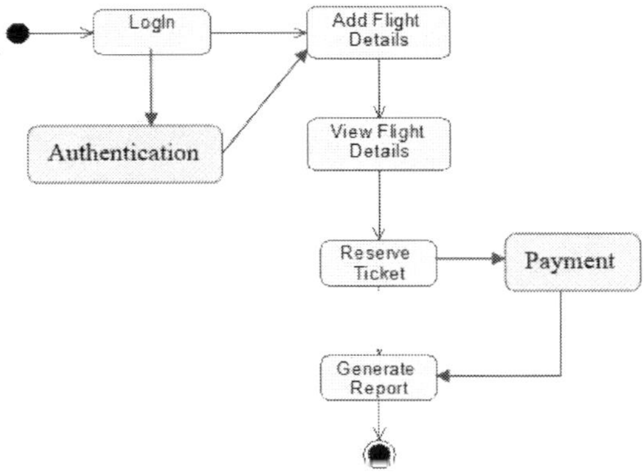

Q.3) Consider a Computer Email System

- Identify actors for **email system**. Explain the relevance of each actor
- One use case is to get email. List four additional use cases at a comparable level of abstraction. Describe each use case with exceptional flow.
- Prepare a **Use case Diagram** for a computer email system. Write **Description for each use case**
- Draw **Class Diagram and Generate Code**.
- Identify at least four states for an email object and draw the **State Transition Diagram** for the same.

Answer 3:

Use Case Diagram

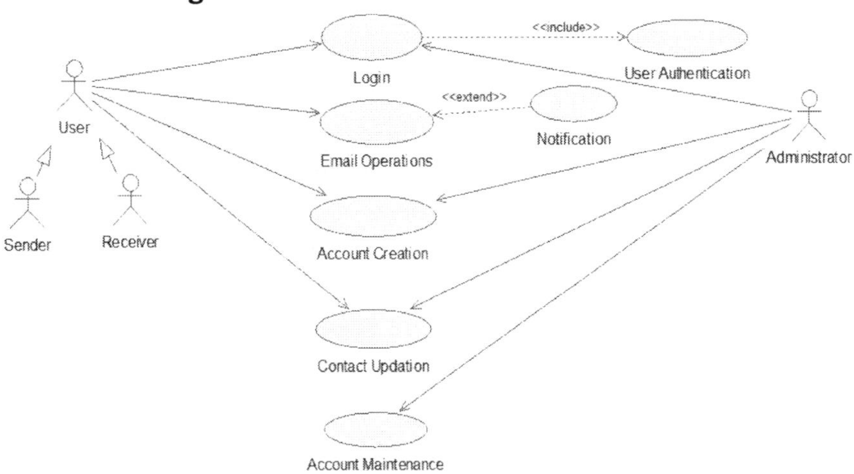

Use Case Descriptions of Major Use Cases

Login Use Case

Brief Description	This use case describes how a user login to the Email System
Actors	User (of generalized types Sender, Receiver) and an administrator
Flow of Events	1. Enter user id 2. Enter password 3. Click on login button
Alternative Flow	If he isn't an existing user then an option to sign up.
Precondition	Sender/Receiver must have an valid account.
Post condition	If the use case was successful the actor is logged in to the system after receiving a success message. If not, the system state is unchanged.

User Authentication Use Case

Brief Description	This use case verifies if the user is an authentic user of the Email System
Actors	User (of generalized types Sender, Receiver) and an administrator
Flow of Events	1. Enter user id 2. Enter password 3. Click on login button
Alternative Flow	If he isn't an existing user then an option to sign up.
Precondition	Sender/Receiver must have entered his login credentials
Post condition	If the use case was successful the actor is logged in to the system after receiving a success message. If not, the system state is unchanged.

Email Operations Use Case

Brief Description	This use case allows the user to perform various options like sending , receiving or browsing mails in the Email System
Actors	User (of generalized types Sender, Receiver)
Flow of Events	1. Create new messages. 2. Send mail 3. Read Existing mails 4. Delete mails
Alternative Flow	If the mailing system is unable to perform the operation, generate a mailer daemon.
Precondition	Sender/Receiver must have entered his login credentials and must be a valid user.

Post condition	If the use case was successful the actor is is able to perform mailing operations. If not, the system state is unchanged.

Account Creation Use Case

Brief Description	This use case enables a new user to create an account in the Email System
Actors	User (of generalized types Sender, Receiver)
Flow of Events	1. Enter basic user details 2. Select user id, password. 3. Perform verification.
Alternative Flow	If the mailing system is unable to perform the operation, the user shall be intimated.
Precondition	Sender/Receiver must have entered his basic details and chosen a password.
Post condition	If the use case was successful the actor will be delivered a success notification. If not, a failure notification shall be raised.

Account Updating Use Case

Brief Description	This use case enables an existing user to modify his account details in the Email System
Actors	User (of generalized types Sender, Receiver)
Flow of Events	1. Enter basic user details 2. Make changes 3. Save state
Alternative Flow	If the mailing system is unable to perform the operation, the user shall be intimated.
Precondition	Sender/Receiver must have entered his basic details and logged into his account.
Post condition	If the use case was successful the actor will be delivered a success notification. If not, a failure notification shall be raised.

Account Maintenance Use Case

Brief Description	This use case enables the administrator to monitor the accounts in the Email System
Actors	System Administrator
Flow of Events	Log into the system Perform maintenance operation Change state
Alternative Flow	If the mailing system is unable to perform the operation, the administrator shall be intimated.

Precondition	Admin must have logged into the system.
Post condition	If the use case was successful the actor will be delivered a success notification and maintenance operation will be performed. If not, a failure notification shall be raised.

Class Diagram of Computer Email System

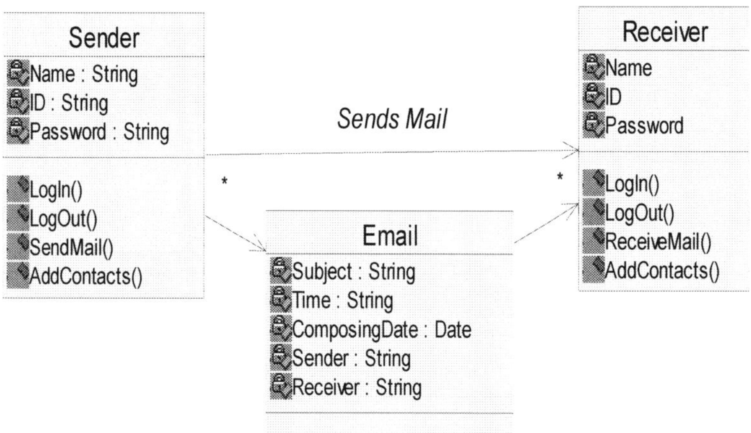

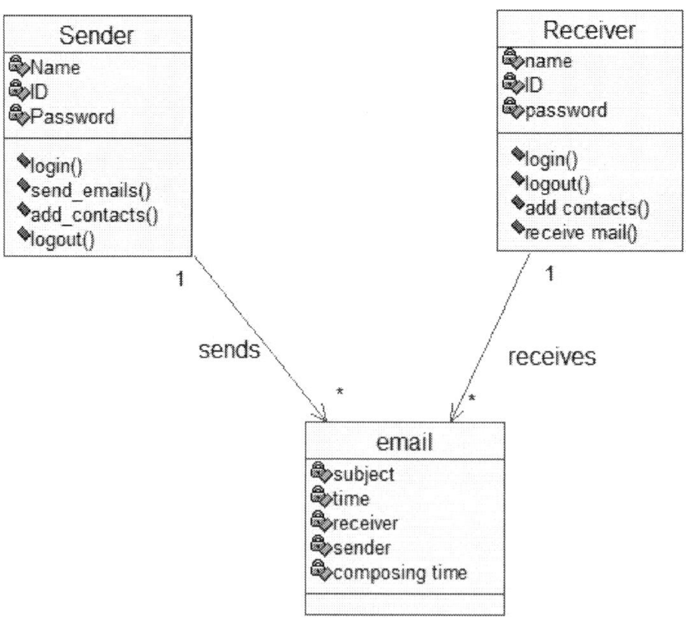

State Diagram of Computer Email System

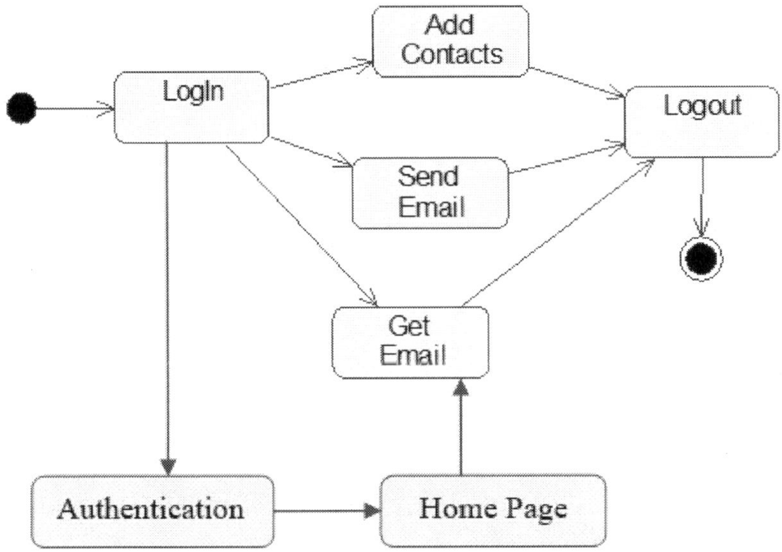

Q.4) Draw Use case Diagram, Class Diagram with Code Generation and Activity Diagram for Student Information System discuss below. Write description of each use case.

System automates the student information system. Student can register for the degree existing in the college and view Course catalogue. It permits the student to register for the course in each semester. Information about each course such as course curriculum, Professor who handles, department that offers course and prerequisite to attend the course will be listed. A student will do 4 courses in each semester on his choice. He must submit with 4 primary and 2 alternative choices of the course. Course will be offered when it is chosen by maximum of 25 students and minimum of 15 students. A course offered by less than the prescribed minimum will be cancelled. The alternative choices are taken into consider in that case. Students change their course within week time of the commencement of the semester. Performance of the student in assignment and continuous internal assessments are recorded electronically and it can be viewed at any time. It also maintains the fee collection details of the student.

Answer 4:
Use Case Diagram

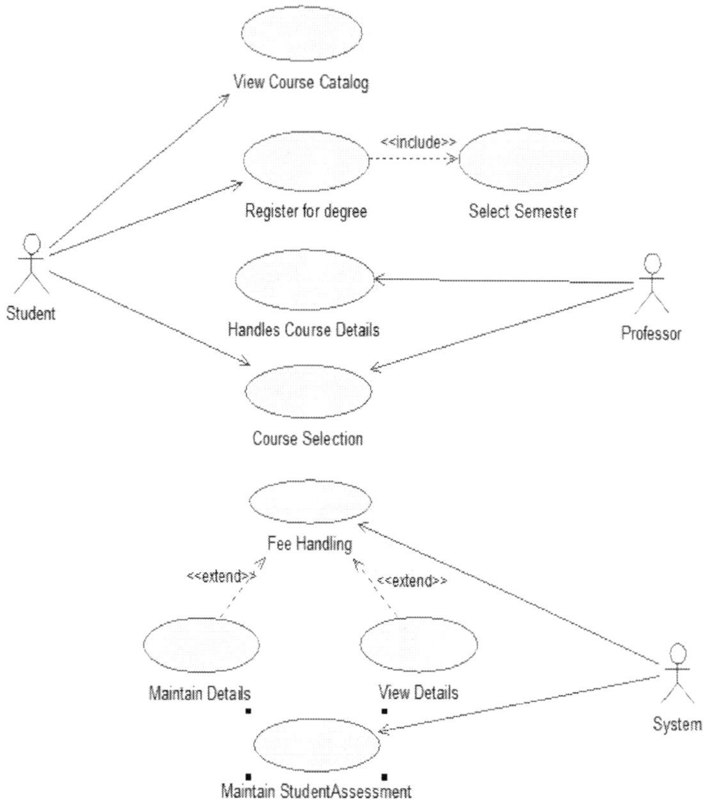

Use Case Descriptions for Major Use cases

View Course Catalog Use Case

Brief Description	This use case enables a student to view course catalog in the Student Information System.
Actors	Student
Flow of Events	1. Log into system 2. Select course 3. View Catalog
Alternative Flow	If the system isn't working, intimate the user.
Precondition	Student must be a registered user.
Post condition	If the use case was successful the student views course catalog. If not, the system state is unchanged.

Register for degree Use Case

Brief Description	This use case enables a student register for a degree in the Student Information System.
Actors	Student
Flow of Events	1. Log into system 2. Register for a degree
Alternative Flow	If the system isn't working, intimate the user.
Precondition	Student must be a registered user.
Post condition	If the use case was successful the student registers for a course. If not, the system state is unchanged.

Select Semester Use Case

Brief Description	This use case enables a student to select a semester for registering for a degree in the Student Information System.
Actors	Student
Flow of Events	1. Log into system 2. Register for a degree 3. Select a semester
Alternative Flow	If the system isn't working, intimate the user.
Precondition	Student must have registered for a degree.
Post condition	If the use case was successful the student selects a course for degree registration. If not, the system state is unchanged.

Handles Course Details Use Case

Brief Description	This use case enables a professor to handle course details in the Student Information System.
Actors	Professor
Flow of Events	1. Log into system 2. Manage Course details
Alternative Flow	If the system isn't working, intimate the user.
Precondition	Professor must have logged into the system.
Post condition	If the use case was successful the professor handles manipulation of course details for degree registration. If not, the system state is unchanged.

Fee Handling Use Case

Brief Description	This use case enables a the system to handle fee details for courses in in the Student Information System.
Actors	system

Flow of Events	1. Select Course 2. Manipulate Fee details
Alternative Flow	If the system isn't working, intimate the user.
Precondition	The course must be registered in the system.
Post condition	If the use case was successful the system manipulates fee details for the system. If not, the system state is unchanged.

Class Diagram Student Information System

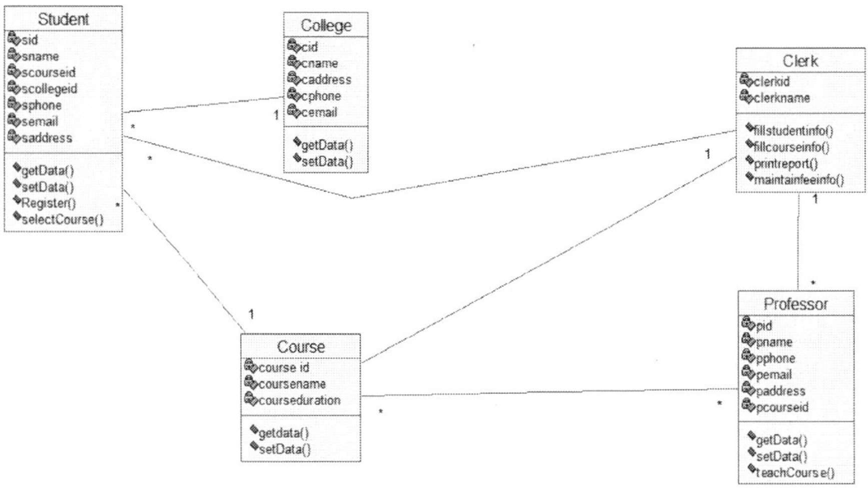

Activity Diagram of student information system

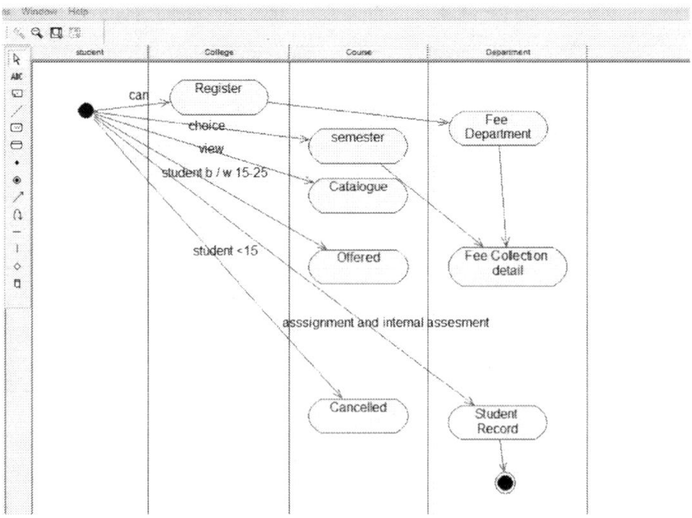

Q.5) Consider a Social Networking Website. The aim of the site is to let people network socially over the Internet. A user can register with the site. On doing so, he is connected with all other registered users of this site. To begin networking, he must search for names through a general search. The system will display the public record of the user under the name. He can select a particular user and invite him to become a friend. On acceptance of the request, the latter's record will be visible in the Friends List of the user and vice versa.

The user can send and receive messages that can be of two types: public and private. A public message, when sent, will be visible to all the registered users browsing the public message list, whereas a private message will be visible only to the recipient. A registered user can upload his photographs and delete previously uploaded photographs. For the persons who do not have any knowledge of this site, an email can be sent, providing information of this site. User can also delete an already added friend from his friend list. He is also allowed to send group messages to a group of friends. Consider only client side functional requirements and answer the followings

Draw **Use case Diagram, and class diagram** using Rational Rose. Write **description for each use case**.

Answer 5:

Use Case Diagram

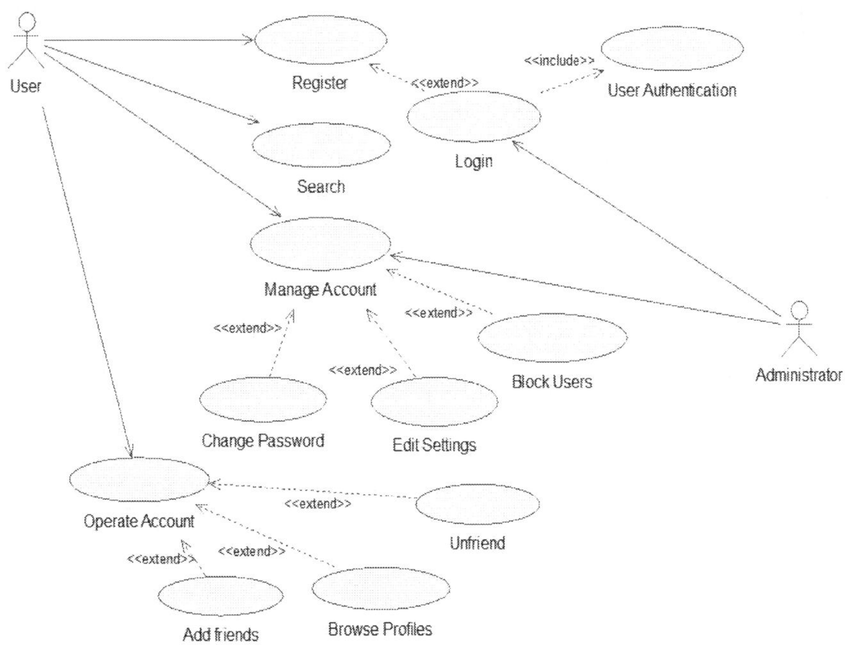

Use Case Descriptions

Login Use Case

Brief Description	This use case describes how a user login to the social networking site.
Actors	User, Administrator
Flow of Events	1. Enter user id 2. Enter password 3. Click on login button
Alternative Flow	If he isn't an existing user then an option to sign up.
Precondition	User/administrator must have entered login credentials.
Post condition	If the use case was successful the actor is logged in to the system after receiving a success message. If not, the system state is unchanged.

User Authentication Use Case

Brief Description	This use case verifies if the user is an authentic user of the social networking website.
Actors	User, Administrator
Flow of Events	1. Enter user id 2. Enter password 3. Click on login button
Alternative Flow	If he isn't an existing user then an option to sign up.
Precondition	User/admin must have entered his login credentials
Post condition	If the use case was successful the actor is logged in to the system after receiving a success message. If not, the system state is unchanged.

Register Use Case

Brief Description	This use case enables a user to register on to the social networking site.
Actors	User,
Flow of Events	1. Enter basic details 2. Select user name 3. Set password 4. Register
Alternative Flow	If the website is down, intimate the user.
Precondition	The site must be operational and working.
Post condition	If the use case was successful the actor is logged in to the system after receiving a success message. If not, the system state is unchanged.

Manage Account Use Case

Brief Description	This use case describes an admin manages user accounts on the social networking site.
Actors	Administrator
Flow of Events	1. Log into the system 2. Perform maintenance activity
Alternative Flow	If he isn't an existing user then an option to sign up.
Precondition	User/administrator must have entered login credentials.
Post condition	If the use case was successful the actor is able to perform maintenance operation on the website. If not, the system state is unchanged.

Search Use Case

Brief Description	This use case describes an a user searches information on the social networking site.
Actors	Users
Flow of Events	1. Log into the system 2. Search for information
Alternative Flow	If the website is down, intimate the user.
Precondition	User/administrator must have entered login credentials.
Post condition	If the use case was successful the actor is able search something on the website. If not, the system state is unchanged.

Class Diagram

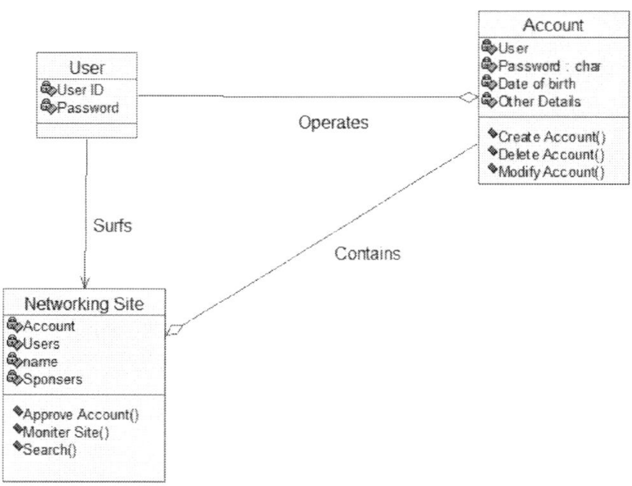

Q.6) Draw **Use case Diagram, Activity Diagram** and **Sequence Diagram (**for each Use case) for **Movie Ticket Machine** problem given below. Write **description of each use case**. Draw **Analysis Model** for **Movie Ticket Machine** problem given below. Draw the **Class Diagram and Generate Code** for the same.

Implement a simple movie ticket vending machine. The movie theater that will use the machine has only one movie and one show time each day. Every morning, the theater manager will turn on the ticket machine, and it will ask him for the name of the movie and the ticket price that day. It will also ask how many seats are in the theater (so it won't sell too many tickets).

When a customer walks up to the ticket machine, he will see the name of the movie, the time, and the ticket price displayed. There is a slot to insert money, a keypad of buttons to enter a number into the "Number of Tickets" field, and a "Buy" button. Printed tickets come out of a slot at the bottom of the machine. Above the ticket slot is a message display (for error messages like "Please enter more money or request fewer tickets" or "SOLD OUT!"). An additional display shows the customer's balance inside the machine. Finally, there is a "Return Change" button so the customer can get his unspent money back.

The ticket machine might look something like this:

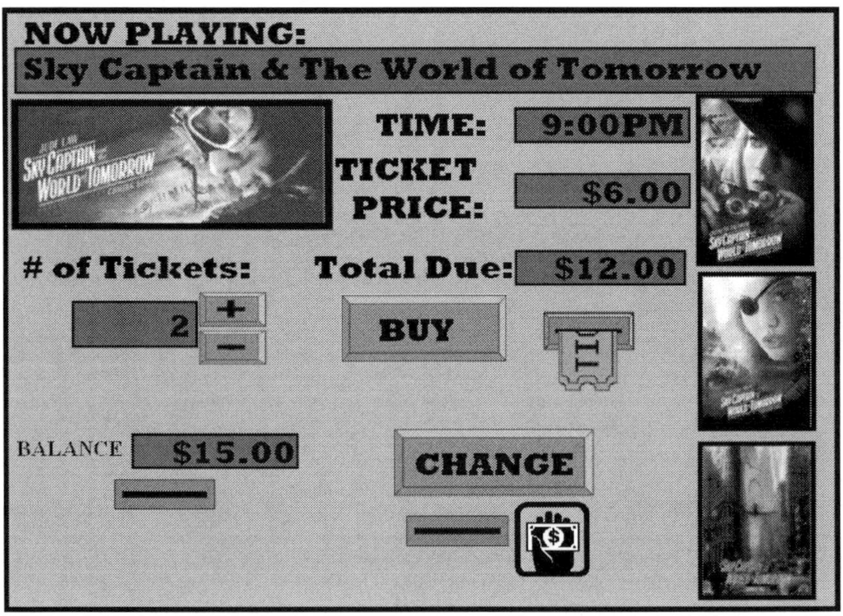

Answer 6:

Use Case

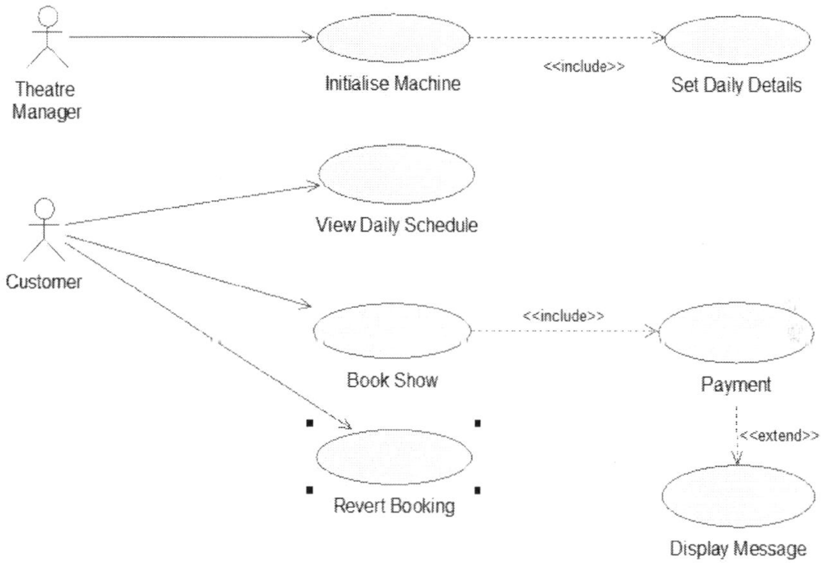

Use Case Descriptions

Initialize Machine Use Case

Brief Description	This use case enables the theatre manager to initialize ticking machine for daily transactions.
Actors	Theatre Manager
Flow of Events	1. Log into ticking system 2. Initialize details
Alternative Flow	If the ticking system is not functioning, it will not start and generate a message.
Precondition	The system is switched on.
Post condition	If the use case was theatre manager will initialize the system. If not, the system state is unchanged.

Set Daily Details Use Case

Brief Description	This use case enables the theatre manager to set daily show ticking machine for daily transactions.
Actors	Theatre Manager
Flow of Events	1. Log into ticking system 2. Initialize details 3. Set daily show details

Alternative Flow	If the ticking system is not functioning, it will not start and generate a message.
Precondition	The system is initialized.
Post condition	If the use case was theatre manager will set up daily show details the system. If not, the system state is unchanged.

View Daily Schedule Use Case

Brief Description	This use case enables the customer to view daily show schedule in the ticking machine.
Actors	Customer
Flow of Events	1. Log into ticking system 2. View daily show details
Alternative Flow	If the ticking system is not functioning, it will not start and generate a message.
Precondition	The system is initialized and daily details recorded.
Post condition	If the use case was customer will be able to view daily show details the system. If not, the system state is unchanged.

Book Show Use Case

Brief Description	This use case enables the customer to book a show in the ticking machine.
Actors	Customer
Flow of Events	1. Log into ticking system 2. View daily show details 3. Book show
Alternative Flow	If the ticking system is not functioning, it will not start and generate a message.
Precondition	The system is initialized and daily details recorded.
Post condition	If the use case was customer will be able to book a show in the system. If not, the system state is unchanged.

Revert Booking Use Case

Brief Description	This use case enables the customer to revert a booked show in the ticking machine.
Actors	Customer
Flow of Events	1. Log into ticking system 2. View daily show details 3. Book show 4. Revert booking

Alternative Flow	If the ticking system is not functioning, it will not start and generate a message.
Precondition	The customer should have booked a show.
Post condition	If the use case was customer will be able to revert a show booking in the system. If not, the system state is unchanged.

Payment Use Case

Brief Description	This use case enables the customer make payment for booking a show in the ticking machine.
Actors	Customer
Flow of Events	1. Log into ticking system 2. View daily show details 3. Book show 4. Payment
Alternative Flow	If the ticking system is not functioning, it will not start and generate a message.
Precondition	The customer should have booked a show.
Post condition	If the use case was customer will be able to make payment for booking a show in the system. If not, the system state is unchanged.

Class Diagram

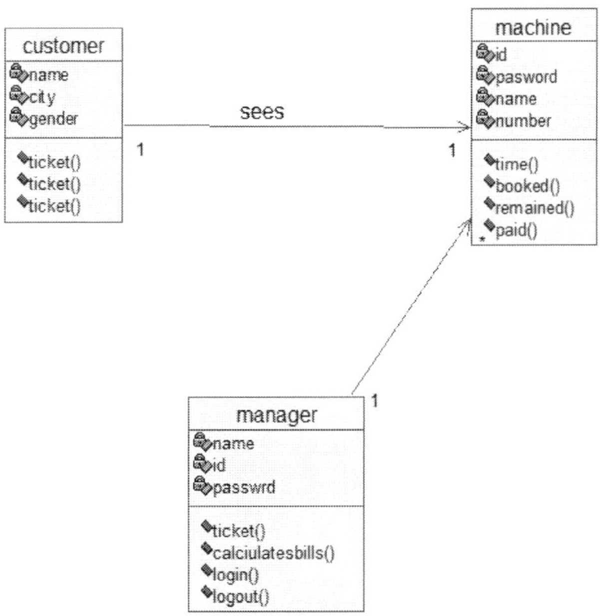

Activity Diagram of Movie Ticket Machine

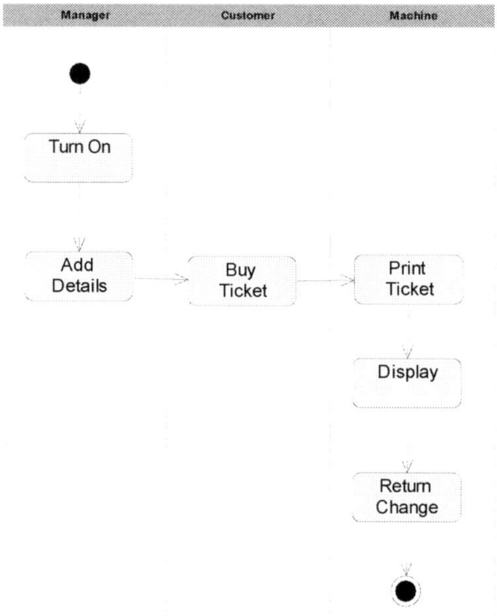

Sequence Diagram of Movie Tick et Machine

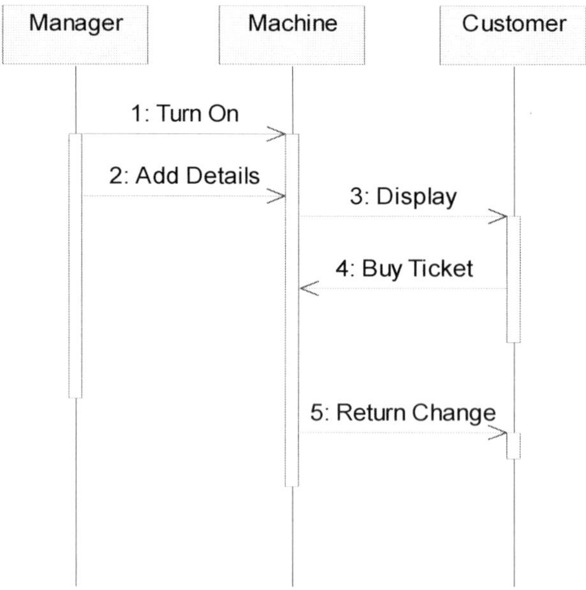

Analysis Model for Movie Ticketing System

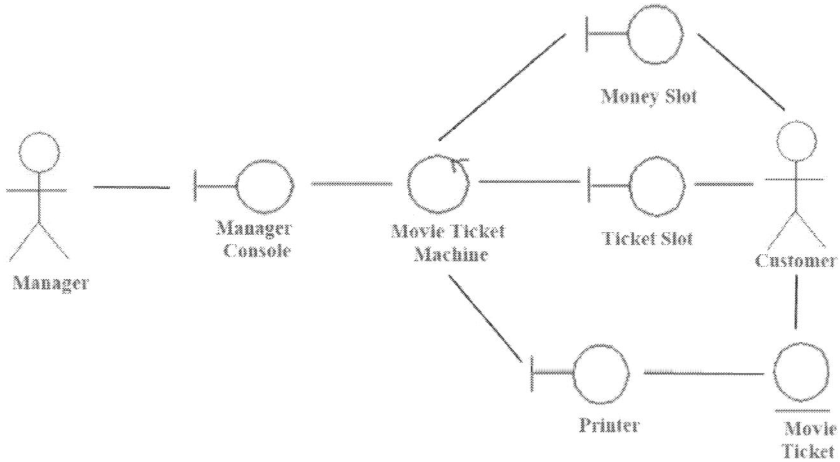

Q.7) Following requirements are for a product purchase system:

- The administrator runs inventory reports. Every time inventory reports are run, inventory data is loaded from disk.
- The administrator updates the inventory stock. Every time inventory stock is updated, inventory data is loaded from disk. Also, every time inventory stock is updated, inventory data is saved to a disk.
- A (general) make-a-sale (hint: meant to be a verb-noun phrase) can be of two specialized kinds: (1) make-a phone order sale; and (2) make-a walk-in sale.
- A sales clerk records the make-a-walk-in sales.
- A telephone operator, a specialized kind of a sales clerk, handles and records all make-a phone orders.
- Whenever there is a make-a-sale, the inventory stock is updated.
- A sale may need to verify a credit card if the purchase is a credit card purchase.
- A sale may need to verify a check if the purchase is a check purchase.

Draw **Use Case Diagram**, including all actors, Use Cases, and relationships. Be sure to use the correct notation for all actors, Use Cases, and relationships. Also be sure to label each and every actor, Use Case, and relationship. Draw **Class Diagram and Generate Code**. Write **description of each use case**.

Answer 7:

Use Case Diagram

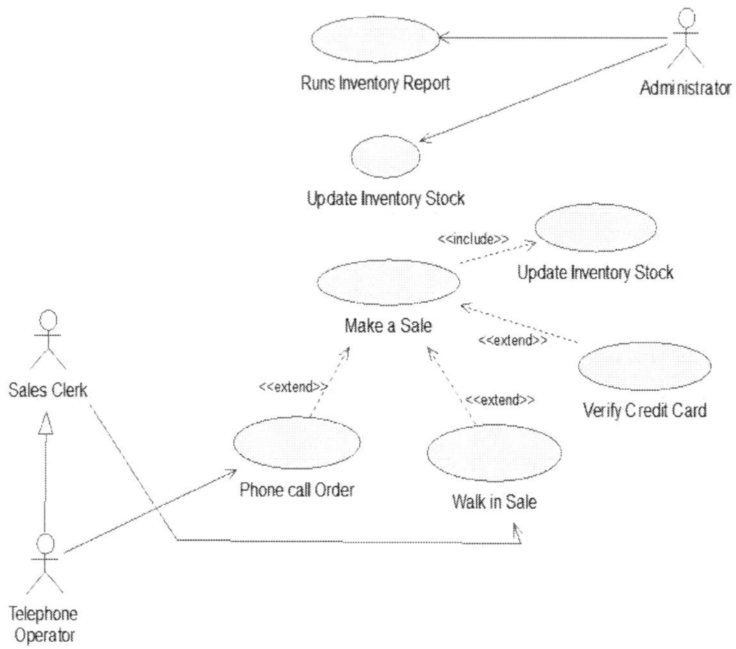

Use Case Descriptions

Run Inventory Report Use Case

Brief Description	This use case enables the administrator to run inventory report in the product purchase system.
Actors	Administrator
Flow of Events	1. Log into the system 2. Run inventory report
Alternative Flow	If he isn't an existing user then an option to sign up.
Precondition	Administrator must have a valid account.
Post condition	If the use case was successful the actor is able to run inventory report. If not, the system state is unchanged.

Update Inventory Stock Use Case

Brief Description	This use case enables the administrator to update inventory stock in the product purchase system.
Actors	Administrator

Flow of Events	1. Log into the system 2. Update inventory stock
Alternative Flow	If the system isn't working, intimate the user.
Precondition	Administrator must have a valid account.
Post condition	If the use case was successful the actor is able to update inventory stock. If not, the system state is unchanged.

Phone Call Order Use Case

Brief Description	This use case enables the the telephone operator to process a phone call order in the product purchase system.
Actors	Telephone operator
Flow of Events	1. Log into the system 2. Attend a phone call 3. Process the order
Alternative Flow	If the system isn't working, intimate the user.
Precondition	A phone call must be recorded in the system.
Post condition	If the use case was successful the actor is able process the phone call order. If not, the system state is unchanged.

Class Diagram

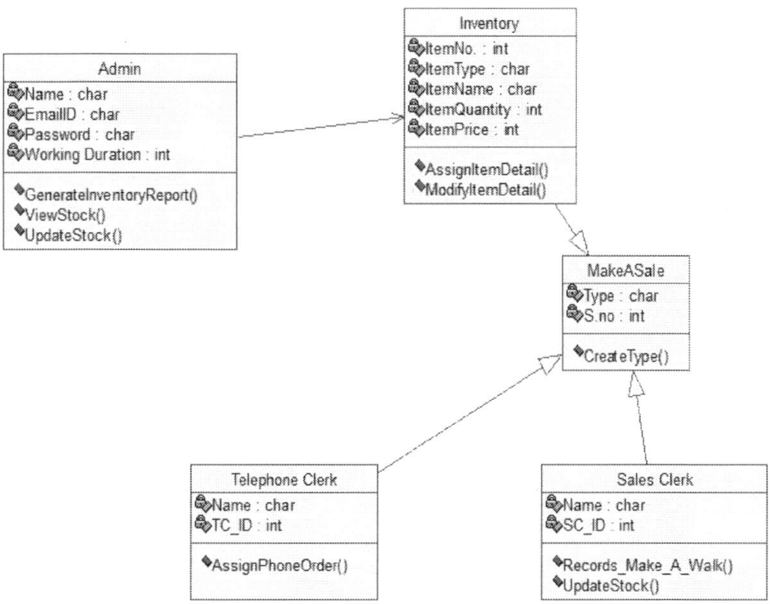

Q.8) Consider a system with which teachers can record and update grades of students. Teachers should also be able to distribute report cards. Here is a complete list of the requirements for the system:

- A teacher can record grades. Whenever grades are recorded, they are also saved to disk.
- A teacher can update grades. Whenever grades are updated, the existing grade is loaded. Then the updated grade is saved to disk.
- A teacher, a registrar, and/or a student can view grades.
- Whenever any of these people view grades, they must always log on to the system. If their log on fails, they must re-authenticate their user name and password.
- A part-time student is a kind of student.
- A registrar can generate report cards.
- A teacher can distribute report cards.

Draw **Use Case diagram**, including all actors, Use Cases, and relationships. Be sure to use the correct notation for all actors, Use Cases, and relationships. Also be sure to label each and every actor, Use Case, and relationship. Draw **Class Diagram and Generate Code**. Write **description of each use case**.

Answer 8:

Use Case Diagram

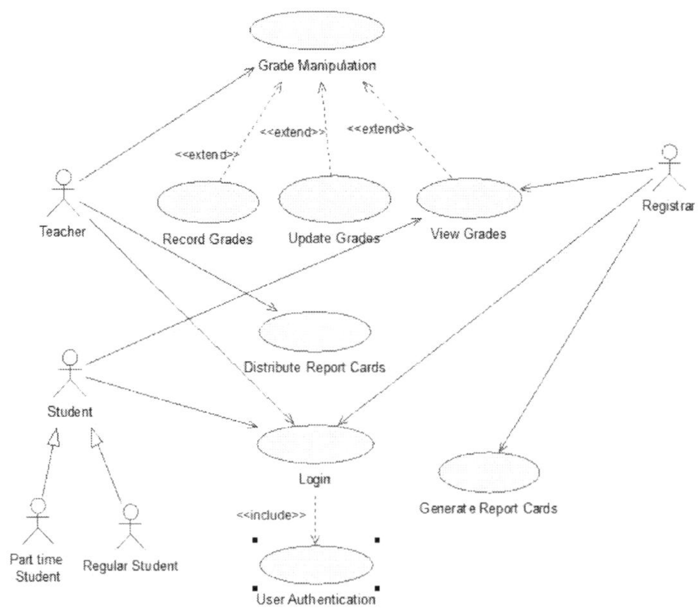

Use Case Descriptions

Grade Manipulation Use Case

Brief Description	This use case enables a teacher to manipulate student grades in his report card.
Actors	Teacher
Flow of Events	1. Select student 2. Manipulate grades
Alternative Flow	If the student hasn't given the exam, inform the teacher.
Precondition	Student must have appeared for the exams.
Post condition	If the use case was successful the teacher successfully manipulates a student's grades. If not, the system state is unchanged.

Record Grade Use Case

Brief Description	This use case enables a teacher to record student grades in his report card.
Actors	Teacher
Flow of Events	1. Select student 2. Review Marks 3. Records grades
Alternative Flow	If the student hasn't given the exam, inform the teacher.
Precondition	Student must have appeared for the exams.
Post condition	If the use case was successful the teacher successfully records a student's grades. If not, the system state is unchanged.

Update Grade Use Case

Brief Description	This use case enables a teacher to update student grades in his report card.
Actors	Teacher
Flow of Events	1. Select student 2. Review Marks 3. Update grades
Alternative Flow	If the student hasn't given the exam, inform the teacher.
Precondition	Student must have appeared for the exams.
Post condition	If the use case was successful the teacher successfully updates a student's grades. If not, the system state is unchanged.

View Grade Use Case

Brief Description	This use case enables a teacher to view student grades in his report card.
Actors	Teacher
Flow of Events	1. Select student 2. Review Marks 3. Records grades 4. View grades
Alternative Flow	If the student hasn't given the exam, inform the teacher.
Precondition	Student must have appeared for the exams.
Post condition	If the use case was successful the teacher successfully view a student's grades. If not, the system state is unchanged.

Distribute Report Card Use Case

Brief Description	This use case enables a teacher to distribute record card of a student.
Actors	Teacher, Student
Flow of Events	1. Select student 2. Select Report Card 3. Distribute card.
Alternative Flow	If the student hasn't given the exam, inform the teacher.
Precondition	Student must have appeared for the exams and his grades should have been recorded.
Post condition	If the use case was successful the teacher successfully distributes a student's report card. If not, the system state is unchanged.

Class Diagram

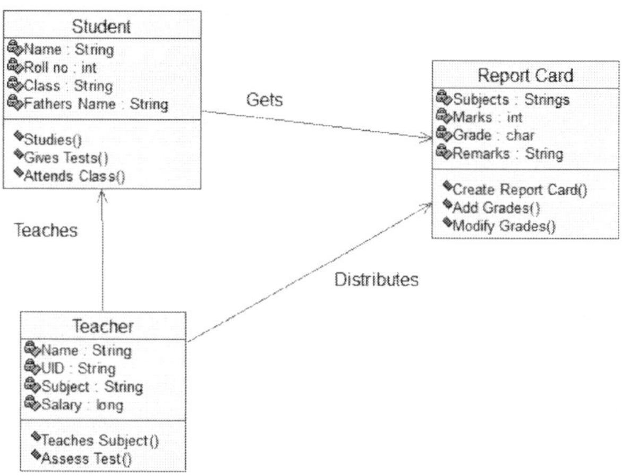

Q. 9) Draw a **Use case Diagram, Activity Diagram, Analysis Model** and **Sequence Diagram (for each use case)**. Write **description of each use case** for a **Simple Answering Machine** given below.

- Use Case: Leave a Message
    - Actor: Caller
    - Pre-Condition: Room on Tape
    - Post-Condition: New Message
- Use Case: Review Messages
    - Primary Path: Review New Messages
    - Alternate Path: No New Messages
- Actor: Owner
    - Use Case: Review Messages Locally
        - Primary Path: Review New Messages
        - Alternate Path: No New Messages
        - Inherit: Review Messages
- Use Case: Review Messages Remotely
    - Primary Path: Review New Messages
    - Alternate Path: No New Messages
    - Inherit: Review Messages
- Uses: Authorize Access
- Use Case: Authorize Access
    - Primary Path: User Authorized
    - Alternate Path: User Not Authorized

Answer 9:

Use Case

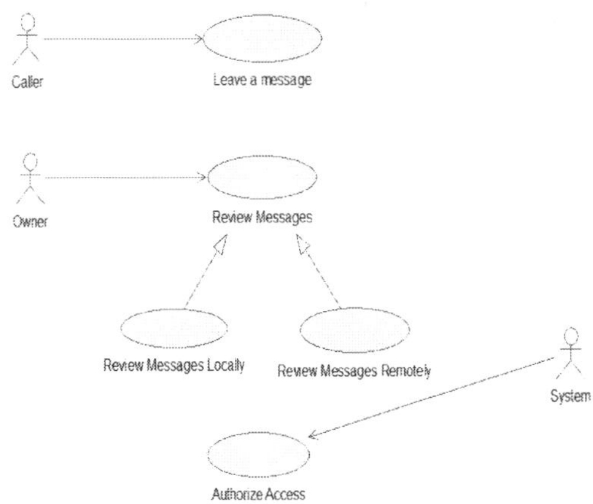

Use Case Descriptions

Leave a message Use Case

Brief Description	This use case enables a user to leave a message in the answering machine.
Actors	Caller to the machine.
Flow of Events	1. Dial number 2. Connect call 3. Leave message
Alternative Flow	If the call doesn't get through, caller shall be intimated.
Precondition	The calling machine should be on Tape
Post condition	If the use case was successful a new message is recorded in the answering machine. If not, the system state is unchanged.

Review message Use Case

Brief Description	This use case enables the owner to hear the messages recorded in the answering machine.
Actors	Owner of the machine.
Flow of Events	1. Pick up the receiver 2. Dial review mode
Alternative Flow	If the machine is not functioning, owner shall be intimated.
Precondition	The calling machine should be on Tape
Post condition	If the use case was successful the owner shall hear them. If not, the system state is unchanged.

Review message locally Use Case

Brief Description	This use case enables the owner to hear the messages locally, recorded in the answering machine.
Actors	Owner of the machine.
Flow of Events	1. Pick up the receiver 2. Dial review mode
Alternative Flow	If the machine is not functioning, owner shall be intimated.
Precondition	The calling machine should be on Tape
Post condition	If the use case was successful the owner shall hear them locally. If not, the system state is unchanged.

Review message remotely Use Case

Brief Description	This use case enables the owner to hear the messages remotely, recorded in the answering machine.
Actors	Owner of the machine.
Flow of Events	1. Pick up the receiver 2. Dial review mode
Alternative Flow	If the machine is not functioning, owner shall be intimated.
Precondition	The calling machine should be on Tape
Post condition	If the use case was successful the owner shall hear them remotely. If not, the system state is unchanged.

Authorize access Use Case

Brief Description	This use case enables the system to authorize access of owner to the answering machine.
Actors	Answering machine system.
Flow of Events	1. User accesses the machine' 2. Waits for authorization
Alternative Flow	If the machine is not functioning, owner shall be intimated.
Precondition	The calling machine should be on Tape
Post condition	If the use case was successful the owner shall be granted access. If not, the system state is unchanged.

Analysis Model of Answering Machine

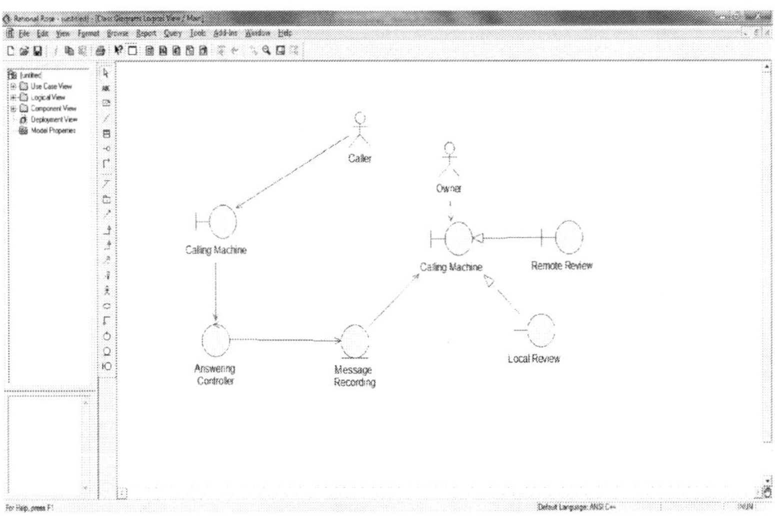

Sequence Diagram of Answering Machine

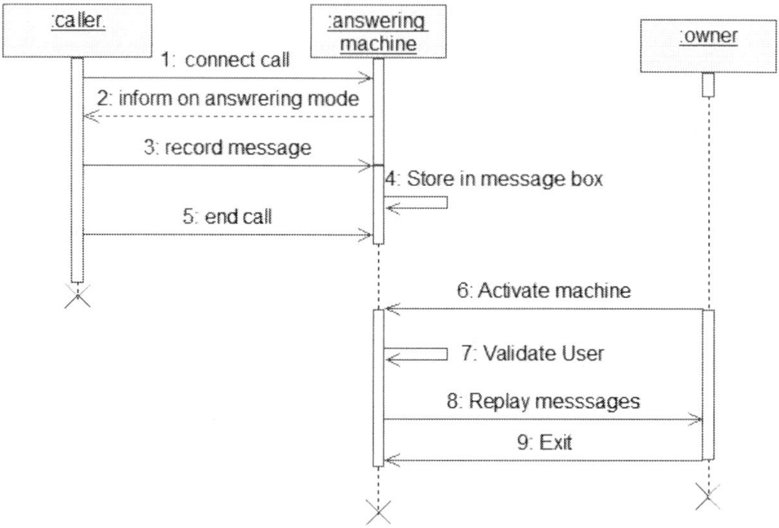

Activity Diagram of Answering Machine

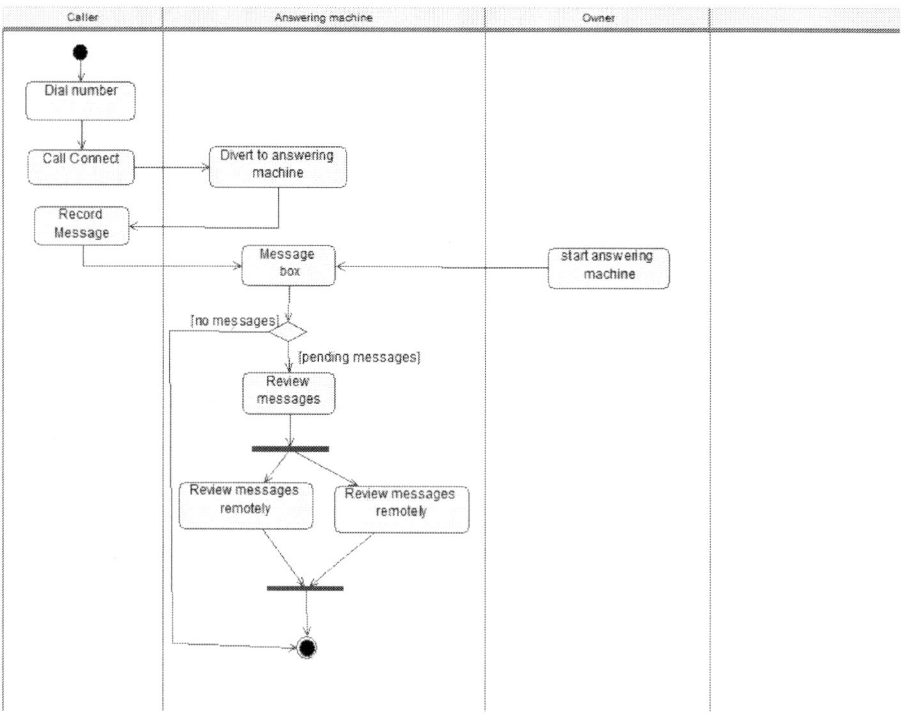

Q. 10) Draw **Use case Diagram, Analysis Model, Class Diagram and Generate Code** for **Health Care Center**. Write **use case description for each use case**

Patient can arrange and cancel appointment with physician using scheduler. Physician decides to Prescribe Medication for Patient. Physician Specifies Drug Info: Medication Name, Dosage Amount, and number Doses & Refills. Computer Cross-Checks for Conflict between Medication and Current Medications/Medical History Prescription Forwarded Electronically to Pharmacy or Else Printed for patient.

Answer 10:

Use Case

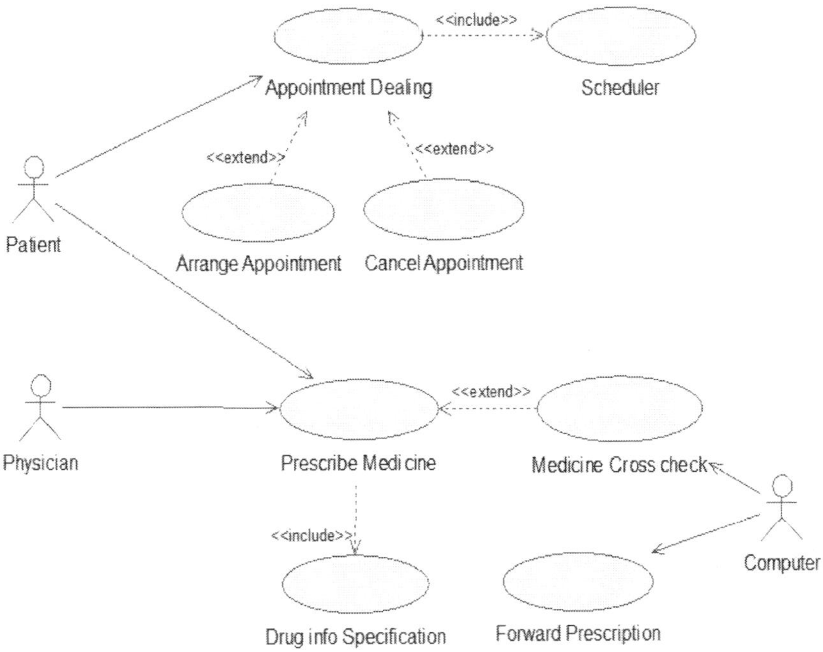

Use Case Descriptions

Appointment Dealing Use Case

Brief Description	This use case enables patient to fix appointments with the health care system.
Actors	Patient
Flow of Events	1. Log into health care system 2. View doctor schedule 3. Fix appointments

Alternative Flow	If the health care system is not functioning, it will not start and generate a message.
Precondition	The physician schedules shall be recorded in the system.
Post condition	If the use case was successful the patient will fix appointment through the system. If not, the system state is unchanged.

Scheduler Use Case

Brief Description	This use case enables system admin to record doctor schedules in the health care system.
Actors	System administrator
Flow of Events	1. Log into health care system 2. Record doctor schedule
Alternative Flow	If the health care system is not functioning, it will not start and generate a message.
Precondition	The system admin is logged into in the system.
Post condition	If the use case was successful the system admin will record schedules into the system. If not, the system state is unchanged.

Arrange Appointment Use Case

Brief Description	This use case enables patient to arrange appointments with the health care system.
Actors	Patient
Flow of Events	1. Log into health care system 2. View doctor schedule 3. Arrange his appointments
Alternative Flow	If the health care system is not functioning, it will not start and generate a message.
Precondition	The physician schedules shall be recorded in the system.
Post condition	If the use case was successful the patient will arrange his appointment through the system. If not, the system state is unchanged.

Cancel Appointment Use Case

Brief Description	This use case enables patient to cancel appointments with the health care system.
Actors	Patient
Flow of Events	1. Log into health care system 2. View doctor schedule 3. cancel his appointments

Alternative Flow	If the health care system is not functioning, it will not start and generate a message.
Precondition	The physician schedules shall be recorded in the system.
Post condition	If the use case was successful the patient will cancel his appointment through the system. If not, the system state is unchanged.

Prescribe Medicine Use Case

Brief Description	This use case enables physician to prescribe medicines to patients through the health care system.
Actors	physician
Flow of Events	1. Log into health care system 2. View patient problem 3. Prescribe medicines
Alternative Flow	If the health care system is not functioning, it will not start and generate a message.
Precondition	The patient details shall be recorded in the system.
Post condition	If the use case was successful the physician will prescribe medicines to the patients. If not, the system state is unchanged.

Medicine Cross Check Use Case

Brief Description	This use case enables computer to cross check prescribed medicines to patients through the health care system.
Actors	System
Flow of Events	1. Log into health care system 2. View prescribed medicines 3. Cross checks medicines
Alternative Flow	If the health care system is not functioning, it will not start and generate a message.
Precondition	The patient details shall be recorded in the system and he should be prescribed a medicine.
Post condition	If the use case was successful the system will cross check availability of prescribed medicines to the patients and generate a confirmation notifictaion. If not, the system state is unchanged.

Class Diagram

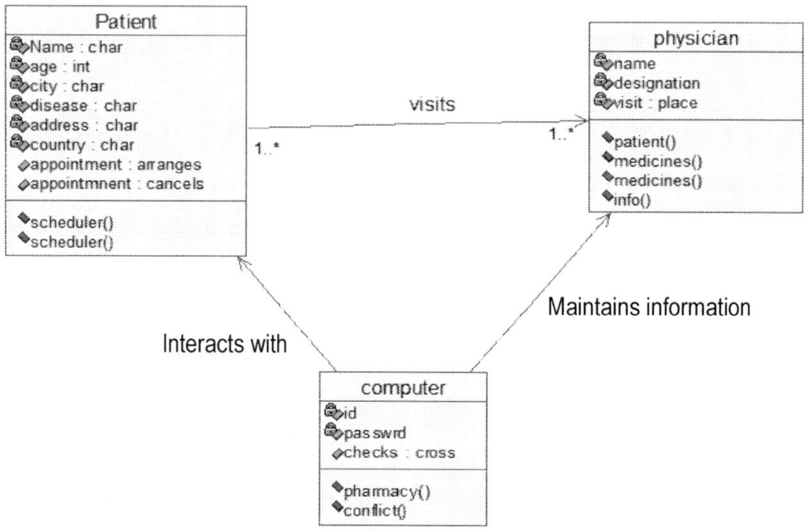

Analysis Model Health Care Center

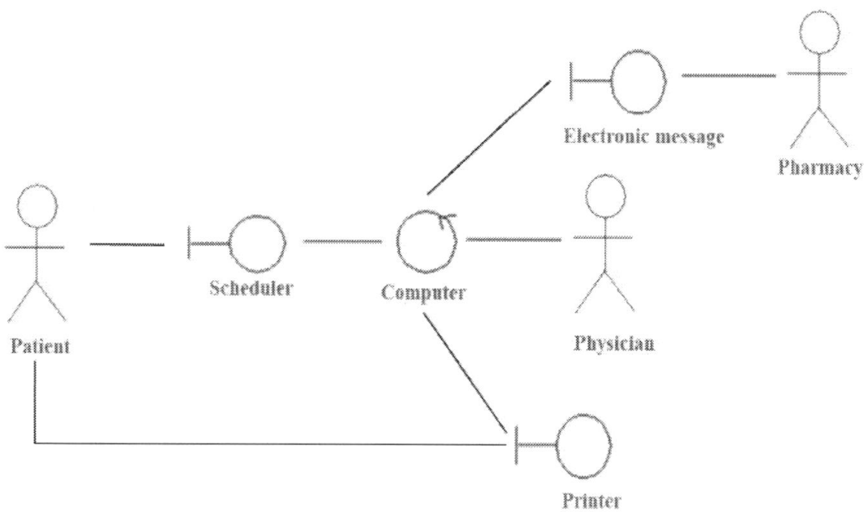

References

[1] Aggarwal K.K., Singh Yogesh, 2008. Software Engineering, 3rd Edition New Age International Publishers.

[2] Ahn Y, Suh J, Kim S, a*nd Kim H, 2003. The Software ma*intenance project effort estimation model based on function points. Journal of Software Maintenance Evolution, Vol. 15, No. 2, pp. 71-85.

[3] Allison Chang, Cynthia Rudin, Michael Cavaretta, Robert Thomas and Gloria Chou, 2013. How to reverse-engineer quality rankings. Mach Learn (2013) 88:369–398 DOI 10.1007/ s10913-013-5295-6, Springer.

[4] Amit Agarwal and kshay Arora, 2017. Security Ecosystem for Data Security and Privacy. Proceedings of 7th International Conference on Cloud Computing, Data Science & Engineering, 12-13 Jan, 2017.

[5] Amit Agarwal and M.M.S. Rauthan, 2017. Remote User Authentication Scheme: A Review. Indian Journal of Science & Technology, 2017.

[6] Amit Agarwal and P. Sirohi, 2015. Computing data storage security framework relating to data integrity, privacy and trust. In Next Generation Computing Technologies (NGCT), 2015 1st International Conference on (pp. 115-118). IEEE, 2015.

[7] Anat Eyal and Tova Milo, 2001. Integrating and customizing heterogeneous e-commerce applications. The VLDB Journal 10: 16–38 (2001) / Digital Object Identifier (DOI) 10.1007/ s007780100045, Springer-Verlag 2001.

[8] Andrea De Lucia, Carmine Gravino, Rocco Oliveto and Genoveffa Tortora, 2009. An experimental comparison of ER and UML class *diagrams for data modeling. Empirical Software*

*Engin*eering (2009) 15:455–492 DOI 10.1007/s10664-009-9127-7, © Springer Science+Business Media, LLC 2009

[9] Andrian *Marcus and Denys Poshyvanyk, 2008. Using th*e conceptual cohesion of classes for fault prediction in object-oriented systems. IEEE Transactions on Software Engineering,VOL.34,NO.2,March/April 2008, pp. 236-278.

[10] B.M. Singh et al.*, 2014. Adaptive binariza*tion of degraded and non-uniformly illuminated documents. International Journal of Document Analysis and Recognition, Springer, DOI 10.1007/s10032-014-0219-6, 2014.

[11] Bache Richard and Mullerburg Monika, 1990. Measures of testability as a basis for quality assurance, Software Engineering Journal, Vol. 23, Issue 11, pp. 295-305, March 1990.

[12] Barbara Poblete an*d Myra Spiliopoulou, 2010. Privacy*-Preserving Query Log Mining for Business Confidentiality Protection. ACM Transactions on the Web, Vol. 4, No. 3, Article 10, Publication date: July 2010, pp. 166-178.

[13] Barbara Poblete and Myra Spiliopoulou, 2010. Privacy-preserving query log mining for business confidentiality protection. ACM Transactions on the Web, Vol. 4, No. 3, Article 10, Publication date: July 2010, pp. 268-285.

[14] Baresi L., Garzotto F., Paolini P., 2000. From web s*ites to web applications: new issues for conce*ptual modeling. ER'2000 Workshop on Conceptual Modeling and the Web, LNCS 2000. Springer-Verlag, 2000, pp. 89-100.

[15] Becker, L.B., Pereira, C.E., Dias, O.P., Teixeira, I.M., Teixeira, J.P.*, 2000. A methodology for automatic object identificati*on from system specification. In: 3rd IEEE International Symposium on Object-Oriented Real-Time Distributed Computing (ISORC 2000), IEEE CS Press (2000), pp. 198–201.

[16] Betty *H.C. Cheng and Enoch Y. Wang, 2002. Formalizing and in*tegrating the dynamic model for object-oriented modeling. IEEE Transactions on Software Engineering, Vol. 28, No. 8, August 2002, pp. 466-484

[17] Bonnell, James P. Davis and Ronald D., 2007. Propositional logic constraint patterns and their use in UML-Based conceptual modeling and analysis. IEEE Tr*ansactions on Knowledge* and Data Engineering, Vol. 19, No. 3 March 2007, pp. 298–309.

[18] Carmine Gravino, Andrea De Lucia, Rocco Oliveto and Genoveffa Tortora, 2010. An experimental comparison of ER and UML class diagrams for data modeling. Empirical Software

Engineering (2010) 15:455–492 DOI 10.1007/s10664-009-9127-10, © Springer Science + Business Media, LLC 2010.

[19] Chen, Z., 1999. Human aspects in object-oriented design: an assessment and a methodology. Behavior and I*nformation Technology, Vol. 11, Issue 5, 11*(5): pp. 256-261.

[20] Chidamber, S.R. and C.F. Kemerer, 1994. A metrics suite for object oriented design. IEEE Transactions on Software Engineering, Vol. 23, Issue 11, *pp. 476-493.*

[21] *Choi Y et al.,* 2009. Identifying sources of opinions with conditional random fields and extraction patterns. In proceedings of Human language technology Conference on Empirical Methods in Natural Language Pr*ocessing (HLT/EMNLP), pp. 355-362,2009.*

[22] *Christian Friberg* and Achim Held, 2005. Support for discretionary role-based access control in acl-oriented operating systems. Proceedings of the Second ACM Role-Based Access Control Work*shop, France, May 20-23, 2005, pp 145-153.*

[23] David Carmel,Vanja Josifovski and Yoelle Maarek,2011.User modeling for web applications. WSDM'11, February 9–12, 2011, Hong Kong, China. ACM 978-1-4503-0493.

[24] Davide *Mazza and Sara Comai, 2012.* A Model-driven methodology to the content layout problem in web applications. ACM Transactions on the Web, Vol. 6, No. 3, Article 10, Publication date: September 2012, pp. 195-208.

[25] D*avid Ferraiolo and Richard* Kuhn, 2004. Role-Based access control. Proceedings of the 15th National Computer Security Conferen*ce, France, May 15-18, 2004,* pp 185-192.

[26] David Scott and Richard Sharp, 2003. Specifying and enforcing application-lev*el web security policies. IEEE Transactions on* Knowledge and Data Engineering, Vol. 15, No. 4, July/August 2003, pp. 195-207.

[27] Davies, S.P., 1998. The structure and content of programming knowledge: disentangling training and language effects in theories of skill development. International Journal of Human-Computer Interaction, Vol. 11, Issue 5, 11(5), pp. 360-380.

[28] Dawson R. Engler, M. Frans Kaashoek and James O'Toole Jr., 2013. Exokernel: operating system architecture for application-level resource management. Proceedings of the 15th ACM Symposium on Operating Systems Principles, France, May 20-23, 2013, pp 145-153.

[29] De Vries, E., 1995. The role of case-based reasoning in architectural design: stretching the design problem space. In

IJCAI'95 Workshop "Reuse of Designs: an Interdisciplinary Cognitive Approach", Chambery, May 20-23, 1993, pp 175-183.

[30] Dorina C. Petriu and Jens Happe, 2014. Introduction to the theme issue on models for quality of software Architec*ture. Software System Model (2014) 13:1237–123*8 DOI 10.1007/s10270-014- 373-1, Springer-Verlag Berlin Heidelberg 2014.

[31] Dr. Mayank Aggarwal, 2011. Simulation of static mobile agent model to prevent denial of service attack using CPNs. International Journal on Recent Trends in Engineering & Technology,Vol 5, No. 01,pp 127-132, Mar 2011. A.C.E.E*.E, USA. ISSN 2158-5563 (Online) ISSN 2158-555*5 (Print).

[32] Elbaum, Sebastian, Alexey G. Malishevsky, and Gregg Rothermel, 2002. Test case prioritization: a family of empirical studies. Software Engineering. IEEE Transactions on the web 28, no. 2 (2002): pp.159-182.

[33] Fernandes, J.M., Machado, R.J., 2001. From use cases to objects: an industrial information systems case study analysis. In: 7th International Conference on Object-Oriented Information Systems (OOIS '01), Springer, Berlin Heidelberg New York (2001), pp. 319–328.

[34] Fernandes, J.M. and Machado, R.J., 2001. From use cases to objects: an industrial information systems case study analysis. In: 7th International Conference on Object-Oriented Information Systems (OOIS '01), Springer, Berlin Heidelberg New York (2001), pp. 319–328.

[35] Ford, J.K. and K. Kraiger, 2007. The application of cognitive constructs and principles to the instructional systems model of training: implications for needs assessment, design and transfer, in International Review of Industrial and Organizational Psychology, C.L. Cooper and I.T. Robertson, Editors. France, May 21-24, 2007, Wiley, pp. 2-46.

[36] G. Gray, 2014. A review of psychometric data analysis and applications in modelling of academic achievement in tertiary education. Journal of Learning Analytics, vol. 1, no. 1, pp. 75–106.

[37] Georgia M. Kapitsaki, Dimitrios A. Kateros, Christos A. Pappas, Nikolaos D. Tselikas and Iakovos S. Venieris, 2008. Model-Driven development of composite web applications. iiWAS2008, November 24–26, 2008, Linz, Austria.2008 ACM 978-1-60558-349 5/08/0011

[38] Georgios Evangelidis and Alexandros Kleftodimos, 2013. An overview of web mining in education. PCI 2013, September 19 - 21 2013, Thessaloniki, Greece Copyright 2013 ACM 978-1-4503-1969-0/13/09.

[39] Giuseppe Ghiani, Federico Bellucci, Fabio Paternò and Claudio Porta, 2012. Automatic reverse engineering of interactive dynamic web applications to support adaptation across platforms. IUI'12, Lisbon, Portugal, February 14-17, 2012, pp. 488–595.

[40] Gray, G., McGuinness, C., and Owende, P., 2013. *An investigation of psychometric measures for modeling academic performance in tertiary education*. In D'Mello, S. K., Calvo, R. A., and Olney, A., editors, Sixth International Conference on *Educational Data Mining, Memphis, Tennessee*, July 6-9, 2013, pages 240-243.

[41] Guindon, R. and B. Curtis, 2009. Control of cognitive processes during software design: what tools are needed? In *Proceedings of ACM CHI'2009: Conference on Human Factors* in Computing Systems. Washington, DC, May 20-23, 2009, pp 145-153.

[42] Gómez J., Cachero C., Pastor O., 2000. Extending a conceptual modeling approach to web application design. In Proc. Conference on Advanced Information Systems Engineering (CAiSE), LNCS 2000, Springer- Verlag, 2000, pp. 79-93.

[43] Gómez J., Cachero C. and Pastor O, 2006. Extending a conceptual modeling approach to web application design. In Proc. Conference on *Advanced Information Systems* Engineering (CAiSE), LNCS 1789, Springer- Verlag, 2006, pp. 79-93.

[44] H. Zhong, T. Xie, L. Zhang, J. Pei, and H. Mei. Mapo, 2009. Mining and recommending API usage patterns. In ECOOP 2009 - Object-Oriented Programming, *23rd European Conference*, France, May 20-23, 2009, pp 318-343.

[45] Hale, J.E., S. Sharpe, and D.P. Hale, 2004. An evaluation of the cognitive processes of programmers involved in software debugging. Journal *of Software Maintenance: Work and Practice*, Vol. 11, Issue 6, 11: pp. 73-91.

[46] Hang Li, Daxin Jiang, J and Ian Pei, 2013. Mining search and browse logs for web search: a survey. ACM Transactions on Intelligent Systems a*nd Technology, Vol. 4, No. 4, A*rticle 57, Publication date: September 2013, pp. 495-508.

[47] Hao-hua Chu and Klara Nahrstedt, 2015. A soft real time scheduling server in UNIX operating system. Proceedings of the European Workshop on Interactive Distributed Multimedia

Systems and Telecommunication Services, Paris, France, August 26-28, 2015, pp 245-253.

[48] Haya Alimagwashi and Steve McIntosh, 2009. Understanding the Government to e-government Transition using a Soft Systems Approach: What is e-government Supposed to do? In Proceedings of the 9th European Conference on government, France, May 10-13, 2009, pp 45-55.

[49] Heena and Ranjna, 2011. A comparative study of UML tools. ACAI '11, July 21 - July 22 2011, Rajpura/Punjab, India Copyright 2011 ACM 978-1-4503-0635-5/11/11.

[50] Hoc, J.-M., 2010. Towards effective computer aids to planning in computer programming: theoretical concern and empirical evidence drawn from assessment of a prototype, in working with computers: theory versus outcomes. Academic: London. p. 215-247.

[51] http://cairo.cs.uiuc.edu/publicati*ons/papers/book-ach*u.pdf

[52] Ibéria Medeiros, Nuno F. Neves and Miguel Correia, 2014. Automatic detection and correction of web application vulnerabilities using data mining to predict false positives. WWW'14, April 7–11, 2014, Seoul, Korea. *ACM 978-1-4503-2744-2/14/14.*http://dx.doi.org/10.1145/2566486.2568024.

[53] Ilka Philippow, Detlef Streitferdt, Matthias Riebisch and Sebastian Naumann, 2005. An approach for reverse engineering of design patterns. Software System Model (2005) 4: 55–70 / Digital O*bject Identifier (DOI) 10.1007/*s10270-004-0059-9, Springer-Verlag 2005.

[54] Ira D, Baxter Michael and Mehlich, 1997. Reverse engineering is reverse forward engineering. Conference on Reverse Engineering, Amsterdam, the Netherlands, Oct*ober 6-8, 1997, pp. 195-208.*

[55] *IvarJac*obson, Magnus Christerson, Patrik Jonsson, and Gunnar Övergaard, 1992. Object-Oriented software engineering: a use case driven approach, Addison-Wesley, Wokingham, England, 1992.

[56] J.J. Jeng and Y. Drissi, 2000. Pens: a predic*tive event notification system for e-commerce envir*onment. In: Proceedings of the 24th Annual International Computer Software and Applications Conference, 2000, COMPSAC, pp. 93-98.

[57] J. Jürjens, 2001. Towards development of secure systems using UMLsec, In H. Hussmann, editor, Fundamental Approaches to Software Engineering, 4th International Conference, Proceedings, LNCS, pages 187–200. Springer, 2001.

[58] J. Jürjens, 2001. Towards development of secure systems using UMLsec. In H. *Hussmann, editor,Fundamental Approaches to S*oftware Engineering, 4th International Conference, Proceedings,LNCS, pages 187–200. Springer, 2001.

[59] James P. Davis, 2007. Propositional logic constraint patterns and their use in uml-based conceptual modeling and analysis. IEEE Transactions on Knowledge and Data Engineering, Vol. 19, No. 3, March 2007, pp. 322-381.

[60] Johnson Rum*baugh, 2006. The Ups and Downs of Object-Oriented Systems Development. Communica*tions of the ACM, Vol. 43, Issue 10, 43(10): pp. 69-73.

[61] John Tuttle Robert, J. Walls Erik, Learned-Miller Brian and Neil Levine, 2010. Reverse engineering for mobile systems forensics with ares. ACM InsiderThreats'10, Chicago, Illinois, USA, October 8, 2010, pp. 488–595.

[62] Jose Brustoloni, Eran Gabber, Abraham Silberschatz and Smith Singh, 2014. Quality of service for legacy applicati*ons. Proceedings of the 9*th International Workshop on Network and Operating Systems Support for Digital Audio and Video, 2014, UK, August 16-19, pp 245-253.

[63] Jose Ignacio Panach, Nelly Condori-Fernández and Francisco Valverde, 2008. Understandability measurement in an early usability evaluation for model-driven development: an empirical study. ESEM'08, October 9–10, 2008, Kaiserslautern, Germany. Copyright 2008 ACM 978-1-59593-971-5/08/10.

[64] Jos Fonseca, Marco Vieira, and Henrique Madeira, 2014. Evaluation of web security mechanisms using vulnerability & attack injection. IEEE Transactions on Dependable and Secure Computing, Vol. 11, No. 5, September/October 2014, pp. 246-291.

[65] Juan Trujillo, 2006. A report on the first international workshop on best practices of UML (BP-UML'06). SIGMOD Record, Vol. 35, No. 3, Sep. 2006, pp. 295-307.

[66] Junior M. F., Fantinato M, and Sun V., 2015. Improvements to the function point analysis method: a systematic literature review. IEEE Transactions on Engineering Management, Vol. 62, No. 4, November 2015, pp. 174-185.

[67] K. Koskimies, T. M¨annist¨o, T. Syst¨a, and J. Tuomi, 1998. Automated support for modeling OO software. IEEE Software transaction, Vol.15, Issue1, January/February 1998, pp. –87 94.

[68] Keng *Siau and Lihyunn Lee, 2004. Are use case and class diagrams comple*mentary in requirements analysis? An experimental study on use case and class diagrams in UML. Requirements Engineering (2004) 9: 229–237, DOI 10.1007/s00766-004-0203-7, Springer-Verlag London Limited 2004.

[69] Kim, J., F.J. Lerch, and H.A. Simon, 2007. Internal representation and rule development in object-oriented design. ACM Tr*ansactions on Computer-Human Inter*action, Vol. 02, Issue 4, 2(4): pp. 357-390.

[70] Kim, J. and F.J. Lerch, 1997. Towards a model of cognitive process in logical design: comparing object-oriented and traditional functional decomposition software methodologies. In Proc. of CHI 92 Human Factors in Computing Systems. Monterey, California, May 20-23, 1997, pp 148-155.

[71] Kölling, M. and J. Ronsenberg, 2005. Guidelines for teaching object orientation with java. In Proc. 6th Annual Conference on Innovation and Technology in Computer Science Education. Canterbury, UK, May 20-23, 2005, pp 145-151.

[72] L. Favre, L. Martínez and C. Pereira, 2000. Transforming UML static models *into object-oriented code. Technology* of Object Oriented Languages and Systems, TOOLS 37 (eds. B. Henderson-Sellers, B. Meyer), IEEE Computer Press, Australia, May 25-28, 2000, pp 185-192.

[73] Lange, B.M. and T. Moher, 2004. Some strategies for reuse in an object-oriented programming environment. In Proc. of CHI'2. Austin, USA: ACM Press, May 24-28, 2004, pp 145-153.

[74] Lionel C. Briand, 2005. An experimental investigation of formality in uml-based development. IEEE Transactions on Software Engineering, Vol. 31, No. 10, October 2005, pp. 366-381.

[75] Luiz A. Ricci and Daniel Schwabe, 2006. An authoring environment for model-driven web applications. WebMedia'06, November 19–22, 2006, Natal, RN, Brazil Copyright 2006 ACM 85-7669-100-0/06/0011.

[76] Luiz A. Ricci and Daniel Schwabe, 2006. An authoring environment for model-driven web applications. WebMedia'06, November 19–22, 2006, Natal, RN, Brazil Copyright 2006 ACM 85-7669-100-0/06/0011.

[77] M. Gogolla and B. Henderson-Sellers, 2002. Formal analysis of UML stereotypes within the uml metamodel. Proceedings of <<UML>> 2002, 5th Int. Conf. Unified Modeling Language

(S. Cook; H. Hussmann; J.M. Jezequel, eds.), Lecture Notes in Computer Science, Springer-Verlag, 2002

[78] Machado, R.J. Fernandes, J.M., Monteiro, P., Rodrigues, H., 2005. Transformation of UML models for service-oriented software architectures. In: Proceedings of 12th IEEE international conference on the engineering of computer based systems (ECBS 2005), June 25-28, 2005, pp. 173–182.

[79] Magdalini Eirinaki and Michalis Vazirgiannis, 2003. Web mining for web personalization. ACM Transactions on Internet Technology, Vol. 3, No. 1, February 2003, pp. 1–27.

[80] Manuel Diaz, Crisian Martin and Bartolome Rubio, 2016. State-of-the-art, challenges and open issues in integration of Internet of things and cloud computing. Elsevier journal of network and computer applications, January- 2016, Vol. 33, Issue 41, pp. 195-207.

[81] Marcin Hernes, Marcin Mleszka, Ngoc Thanh Nguyen and Andrzej Bytniewski, 2015. The automatic summarization of text documents in the cognitive integrated management information system, IEEE Conference on Computer Science and Information Systems (FedCSIS), 2015, pp. 1387-1396.

[82] Marco Brambilla, Stefano Ceri, and Piero Fraternali, 2006. Process modeling in web applications. ACM Transactions on Software Engineering and Methodology, Vol. 15, No. 4, October 2006, pp. 155-188.

[83] Marco Brambilla, Stefano Ceri, and Piero Fraternali, 2006. Process modeling in web applications. ACM Transactions on Software Engineering and Methodology, Vol. 15, No. 4, October 2006, pp. 195-208.

[84] Marco Vieira, Jose Fonseca and Henrique Madeira, 2014. Evaluation of web security mechanisms using vulnerability & attack injection. IEEE Transactions on Dependable and Secure Computing, Vol. 11, No. 5, September/October 2014, pp. 295-309.

[85] Martin F. Arlitt and Carey L. Williamson, 2013. Internet web servers: workload characterization and performance implications. IEEE Transactions on Networking, Vol.45, Issue 30, pp. 875-887.

[86] Michel R. V. Chaudron, Werner Heijstek and Ariadi Nugroho, 2012. How effective is UML modeling? An empirical perspective on costs and benefits. Software System Model (2012) 11:571–580 DOI 10.1007/s10270-012-0278-4, Springer-Verlag 2012.

[87] Minaei-Bidgoli, Behrouz, and William F. Punch, 2003. Using genetic algorithms for data mining optimization in an educational web-based system. Genetic and evolutionary computation conference. Springer Berlin Heidelberg, 2003.

[88] Myriam Ben Saad and Stéphane Gançarski, 2011. Archiving the web using page changes patterns: a case study. JCDL'11, June 13–17, 2011, Ottawa, Ontario, Canada. Copyright 2011 ACM 978-1-4503-0744-4/11/06.

[89] Nasib S. *Gill and Balkishan, 2008. Dependency a*nd Interaction Oriented Complexity Metrics of Component-Based Systems. ACM SIGSOFT, Volume 33, Number 2, 2008, ACM Press, New York, NY, USA.

[90] Nasib S. Gill and Tomar, 2010. Modified development *process of component-based software engineering. A*CM SIGSOFT, Vol. 35, No. 2, pp. 1-6 New York, USA.

[91] Naveen Kumar, 2012. User Authentication Using Musical Password. Published in the International Journal of Computer Applications, Vo*lume-59, pp 45-51, December 2012.*

[92] *Naveen Kumar and M* N Doja, 2006. Secure Authentication Using Graphical Password. In the proceeding of International conference on Web engineering and Applications, Mac Millian Publication, pp 251-262, 2006.

[93] Naveen Kumar and M N Doja, 2007. Comparative Analysis on Alternative Schemes of User Authentication. In proceeding of second international Conference on Embedded System, Mobile Communication and Computing, Banglore, India, pp. 168-179, 2007.

[94] Nitish Pathak, Girish Sharma and B. M. Singh, 2014. Forward engineering based implementation of TOS in social networking. International Journal of Computer Applications (IJCA), Vol. 102, No.11, September-2014, pp: 33-38, ISSN: 0975 - 8887. Foundation of Computer Science, New York, USA.

[95] Nitish Pathak, Girish Sharma and B. M. Singh, 2015. Ex*perimental analysis of SPF* based secure web application. International Journal of Modern Education and Computer Science (IJMECS), ISSN: 2075-0161 (Print), ISSN: 2075-017X (Online), Volume 7, No. 2, February 2015, pp 48-55, MECS, Hong Kong.

[96] Nitish Pathak, Girish Sharma and B. M. Singh, 2015. Towards designing of SPF based secure web application using UML 2.0. International Journal of Systems Assurance Engineering

and Management, DOI: 10.1007/s13198-015-0338-6, 28th January, 2015, Springer.

[97] Nitish Pathak, Girish Sharma and B. M. Singh, 2017. An empirical perspective of roundtrip engineeri*ng for the development of secure web application using* UML 2.0. International Journal of Intelligent Systems and Applications, Vol.9, No.06, 2017, pp 1-15, Hong Kong.

[98] Nitish Pathak, Girish Sharma and B. M. Singh, 2017. *UML 2.0 based framework for the develo*pment of secure web application. International Journal of Information Technology (IJIT), DOI: 10.1007/s141870-017-0001-3, 22nd February, 2017, Springer.

[99] O. R Marban, J. Segovia, 2013. Extending UML for modeling data mining projects (DM UML). Information Technology Software Engineering, vol.3 Issue-2, pp.1-17, 2013.

[100] Oscar Pastor, Silvia Abrahão and Joan Fons, 2009. Building e-commerce applications from object-oriented conceptual models. ACM Inc., 1515 Broadway, New York, NY, USA.

[101] P. Epstein and R. Sandhu, 1999. Towards a UML based approach to role engineering. In Proceedings of the fourth ACM workshop on Role-based access control, ACM Press, 1999, pages 135–143.

[102] P. Epstein and R. Sandhu, 1999. Towards a UML based approach to role engineering. In Proceedings of the fourth ACM Workshop on Role-based access control, pages 135–143. ACM Press, 1999.

[103] Pedro Valderas and Vicente Pelechano, 2011. A survey of requirements specification in model-driven development of web applications. ACM Transactions on the Web, Vol. 5, No. 2, Article 10, Publication date: May 2011, pp. 566-581.

[104] Per Runeson and Martin Höst, 2009.Guidelines for conducting and reporting case study work i*n software engineering*. Empirical Software Engineering. (2009) 14:131–164 DOI 10.1007/s10664-009- 102-8, open access at Springerlink. com, December 2009.

[105] Peter Desnoyers, Timothy Wood and Prashant Shenoy, 2012. Modellus: automated modeling of complex internet data center applications. ACM Transactions on the Web, Vol. 6, No. 2, Article 8, Publication date: May 2012, pp. 676-695.

[106] Pfleeger, S.L., 1995. Experimental desig*n and analysis in software engineering, part 3: types of* experimental design. ACM SIGSOFT Software Engineering Notes, 1995. 20(2): pp. 1*4-16.*

[107] *Pfleeger, S.L.,* 1995. Experimental design and analysis in software engineering, part 4: choosing an experimental design. ACM SIGSOFT Software Engineering Notes, 19*95. 20(3): p. 13-15.*

[108] *Phyoung* Jung Kim and Young Ju Noh, 2013. Mobile agent system architecture for supporting mobile market application service in mobile computing environment. IEEE International Conference on Geometric Modeling and Graphics, China, May 20-23, 2013, pp. 149-153.

[109] Piotr Kosiuczenko, 2009. Redesign of UML class diagrams: a formal approach. Software System Model (2009) *8:165–183 DOI 10.1007/s10270-007-0068-6, November 2007 © Springer*-Verlag 2009.

[110] Power, J. F. and Malloy, 2000. An approach for modeling the name lookup problem in the C + + programming language. In ACM Symposium on Applied Computing, Como, Italy, July15-18, 2000, pp 388-396.

[111] Pressman, R.S., 1997. *Software Engineering: A Pra*ctitioners Approach. 4th ed., McGraw-Hill.

[112] Przemyslaw Pardyak, Craig Chambers, Emin Gun Sirer and Susan Eggers, 2011. Extensibility, safety and performance in the SPIN operating system. Proceedings of the 15th ACM Symposium on Operating Systems Principles 2011, Canterbury, UK, June 6-9, pp 145-153.

[113] R. Breu, R. Grosu, F. Huber, B.Rumpe and W. Schwerin, 1997. Towards a precise semantics for object-oriented modeling techniq*ues. Proceedings of the ECOOP'97*, Lecture Notes in Computer Science 1241, Springer-Verlag, 1997, pp. 314-364.

[114] R. Fletcher and A. S. M. Sajeev, 1996. A framework for testing object-oriented software using formal specifications. In A. Strohmeier, editor, Reliable Software Technologies, number 1088 in Lecture Notes in Computer Science, pp 159-170. Springer, 1996.

[115] R.W. Selby and V.R. Basili, 1998. Clean room software development: an empirical evaluation. IEEE Transaction on Software Engineering, vol. 13, no. 9, Sept. 1998, pp. 1027-1037.

[116] Rajendra Akerkar, Nick Bassiliades and John Davies, 2014. Work and applications in web intelligence, mining, and semantics. WIMS '14, June 02 - 04 2014, Thessaloniki, Greece, Copyright 2014 ACM 978-1-4503-2538-7/14/06.

[117] Rist, R.S., R.L. Winder and D.J. Gilmore, 2005. Search through

multiple representations, in user-centred requirements for software engineering environments. Editors. Springer-Verlag: Berlin. p. 165-176.

[118] Ronan Barrett, Claus Pahl, Lucian M. Patcas and John Murphy, 2006. Model driven distribution pattern design for dynamic web service compositions. ICWE'06, July 1114, 2006, Palo Alto, California, USA. ACM 1595933522/06/0007.

[119] Rosson, M.B. and J.M. Carroll, 1996. The reuse of uses in smalltalk programming. ACM Transactions on Computer-Human Interaction, Vol. 23, Issue 41, 3(3): p. 219-253.

[120] Rosson, M.B. and J.M. Carroll, 1997. Active programming strategies in reuse. In Proc. of ECOOP'93.Kaiserlautern, Germany: Springer-Verlag, August 14-17, 1997, pp 165 173.

[121] Roy A, Karforma S, 2011. A survey on e-governance security. International Journal of Computer Engineering and Computer Applications (IJCECA). Fall Edition 2011, Vol 08 Issue No. 01, pp: 50-62, ISSN 0974-4983.

[122] Rumbaugh, J., I. Jacobson, and G. Booch, 1999. The Unified Modeling Language Reference Manual. Reading, Massachusetts, Addison-Wesley.

[123] S. Kotsiantis, C. Pierrakeas, and P. Pintelas, 2004. Predicting students' performance in distance learning u*sing machine learning techniques. Applied Artifici*al Intelligence, vol. 18, no. 5, May 2004, pp. 411–426.

[124] S. M. K. Quadri and Ahmed Peer, 2012. Scaling Up for the Streaming Data. International Journal of Engineering and Advanced Technology (IJEAT). Volume-2, Issue-2, pp: 362-368, December 2012.

[125] S. M. K Quadri, Shabia Shabir Khan and Mushtaq Ahmed Peer, 2012. Scaling Up for the Streaming Data. International Journal of Engineering and Advanced Technology (IJEAT), Volume-2, Issue-2, pp: 362-368, December 2012.

[126] S. M. K Quadri *and Sheikh Umar Farooq, 2011. 3W's of Static Softwar*e Testing Techniques. Global Journal of Computer Science and Technology (USA), Volume 11 Issue 6 Version 1.0, pp. 77-86, April, 2011.

[127] S. Tanenbaum, 2*001. Modern Operating Systems. Prentice Hall.*

[128] *Salvador Trujillo, Don B*atory and Oscar Diaz, 2007. Feature oriented model driven development: a case study for portlets. 29th International Conference on Softwa*re Engineering (ICSE'07), IEEE 0-7695-2828-7/07 $20.00* © 2007

[129] Sara Comai and Davide Mazza, 2012. A model-driven methodology to the content layout problem in web applications. ACM Transactions on the Web, Vol. 6, No. 3, Article *10, Publication date: September 2012, pp. 165-195.*

[130] *Sara Comai and* Davide Mazza, 2012. A model-driven methodology to the content layout problem in web applications. ACM Transactions on the Web, Vol. 6, No. 3, Article 10, Publication date: September 2012, pp. 266-281.

[131] Shadab Khan, Aruna B Dulloo and Meghna Verma, 2014. Systematic review of requirement elicitation techniques. International Journal of Information and Computation Technology. ISSN 0974-2239, International Work *Publica*tions House, Volume 4, Number 2 (2014), pp. 133-138.

[132] Shwetank Arya and R. P. Agarwal, 2011.Data Mining Technique for Agriculture and Related Areas. Internati*onal Journa*l of Advanced Work in Computer Science,Volume 2, No. 6, pp. 43-45, ISSN:0976–5697.

[133] Shwetank Arya and R. P. Agarwal, 2012. Constraint Based Data Mining Focusing Farming case stud*y. Iternational Journal of Computer Appli*cation (IJCA), Volume 59, No. 12, pp. 8-11, December 2012 Published by Foundation of Computer Science, New York, USA, ISSN: 0975–8887.

[134] Simona Bernardi, Jos´E *Merseguer and Dorina C. Petriu, 2012. Depen*dability modeling and analysis of software systems specified with UML. ACM Computing Surveys, Vol. 45, No. 1, Article 2, Publication date: November 2012, pp. 390-398.

[135] Simona Bernardi, Jos´E Merseguer and Dorina C. Petriu, 2012. Dependability modeling and analysis of software systems specified with UML. ACM Computing Surveys, Vol. 45, No. 1, Article 2, Publication date: November 2012, pp. 566-581.

[136] Soloway, E. and K. Ehrlich, 1984. Empirical studies of programming knowledge. IEEE transactions on software engineering. Vol. 10, Issue 05, 10(5), pp. 595-609.

[137] Stefanowski and Jerzy, 2005. An experimental study of methods combining multiple classifiers-diversified both by feature selection and bootstrap sampling. Issues in the Representation and Processing of Uncertain and Imprecise Information, pp. 337-354.

[138] Stroustrup, B., 1991. The C++ Programming Language. Reading, Massachusetts, *Addison-Wesley.*

[139] *Sylvia Osborn, Ravi S*andhu, and Qamar Munawer, 2008. Configuring role-based access control to enforce mandatory and discretionary access control policies. ACM Transactio*ns on Information and Systems Security, Vol. 23, Issue* **1**1, pp. 195-207.

[140] T. B. Lee and M. Fischetti, 1999. Weaving the web: the original design and ultimate destiny of the World Wide Web by its inventor. HarperCollins, United States of Americ*a, September 1999, pp. 177-195.*

[141] *Tamar Richner and St´ephane Ducasse,* **19**99. Recovering high-level views of object-oriented **a**pplications from static and dynamic information. In Hongji Yang and Lee White, editors, Proceedings ICSM'99 (International Conference on Software Maintenance). IEEE, Septemb*er 1999.*

[142] *Tamar Richner and St´ephane Ducasse, 1999. Recoveri*ng high-level views of object-orient**ed ap**plications from static and dynamic information. In Hongji Yang and Lee White, editors, Proceedings ICSM'99 (Inter*national Conference on Software Maintenance). IEEE, September 1999.*

[143] *Thi*ago Jabur Bittar, Renata P. M. Fortes and Luanna Lopes Lobato, 2009.Web communication and interaction modeling using model-driven development. SIGDOC'09, October 5–7, 2009, Bloomington, Indiana, USA. Copyright 2009 ACM 978-1-60558-559-8/09/10.

[144] Thiago Jabur Bittar, Renata P. M. Fortes and Luanna Lopes Lobato, 2009. Web communication and interaction modeling using model-driven development. SIGDOC'09, October 5–7, 2009, Bloomington, Indiana, USA. Copyright 2009 ACM 978-1-60558-559-8/09/10.

[145] Vinod Sharma, G.S Sambyal and V. Mansotra, 2009. Digital Divide in education – A concern for J&K State, INDIACom-2009 ISSN 0973 – 7529, ISBN 978-81-904526-2-5.

[146] Vinod Sharma et al., 2008. Critical Analysis of Software Usage vs. Cost. 2nd National Conference; INDIACom – 2008 computing For Nation Development, February 08 – 09, 2008, pp.1-6.

[147] Vinod Sharma et al., 2009. A Model for Electronic Health Communication in J&K State in India. 2nd IEEE International Conference on Computer Science and Information Technology, 2009. ICCSIT 2009. DOI: 10.1109/ICCSIT.2009.5234376 · Source: IEEE Xplore

[148] Vinod Sharma et al., 2009. IT initiatives in education 'A case study for J&K State'. 2nd IEEE International Conference on Computer Science and Information Technology, 2009. ICCSIT 2009. DOI: 10.1109/ICCSIT.2009.5234382 · Source: IEEE Xplore

[149] W. Ahrendt, T. Baar, B. Beckert, M. Giese, R. Hähnle, W. Menzel, W.Mostowski and P. Schmitt, 2002. The key system: integrating object-oriented design and formal methods. Proceedings of FASE 2002 ETAPS 02, Grenoble, France. Available at http://i12www.ira.uka.de/~projekt/index.html, 2002.

[150] W. Zheng, Q. Zhang, M. R. Lyu, and T. Xie, 2010. Random unit-test generation with mut-aware sequence recommendation. In ASE 2010, 25th IEEE/ACM International Conference on Automated Software Engineering, Antwerp, Belgium, September 20-24, 2010, pp 293-296.

[151] Wagner, 2006. Building Semantic Webs for e-government with Wiki technology. In Electronic Government, an International Journal, Inderscience Publishers, Volume 3, Number 1, pp.36-55.

[152] Wasserman, Stanley, and Katherine Faust, 1994. Social network analysis: methods and applications. Vol. 8, Cambridge university press, 1994, pp. 1–7.

[153] Wim De Pauw, Richard Helm, Doug Kimelman, and John Vlissides, 1993. Visualizing the behavior of object-oriented systems. In Proceedings OOPSLA '93, pages. ACM Press, May 15-18, 1993, pp 326–337.

[154] X. Jiang and X. Qian, 2009. Study on intelligent e-shopping system based on data mining. School of Electronic Information & Electrical Engineering Changzhou Institute of Technology CZU, Chanzhou, China, 2009, pp. 1–7.

[155] Y. Labiche, 2014. Integration testing object-oriented software systems: an experiment driven work approach. IEEE CCECE 2014 Niagara Falls, Canada, pp. 000652-000655.

[156] Y. Shen, L. Xing and Y. Peng, 2007. Study and application of web-based data mining in e-business. In: Proceedings of 8th ACIS International Conference on Software Engineering, Artificial Intelligence, Networking and Parallel / Distributed Computing, France, May 20-23, 2007, pp 145-153.

Appendices

APPENDIX I

Round trip engineering and Code of the Example:

As per the given case study, Fig. I-A shows the Generic Use case diagram of hospital and medical sector:-

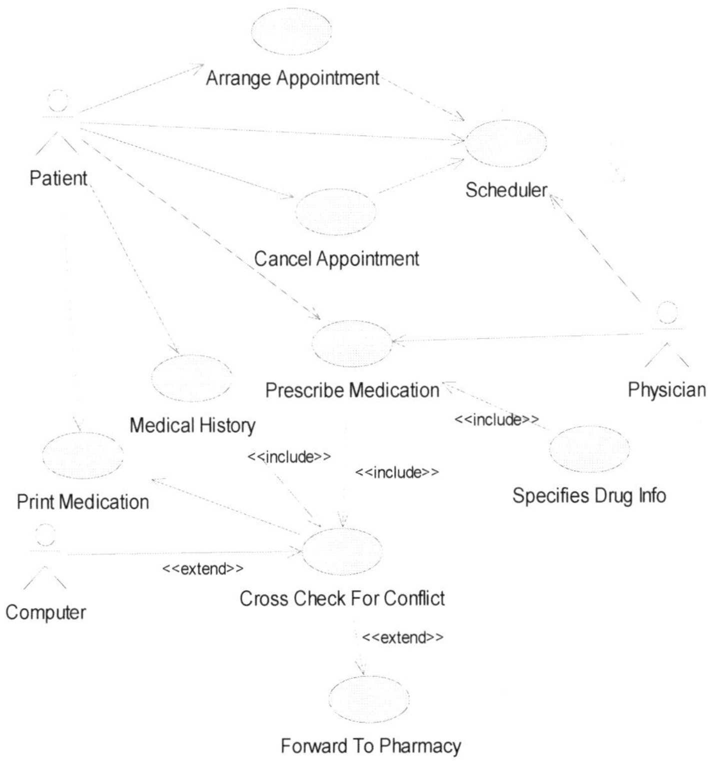

Fig. I-A. Use case diagram of hospital and medical sector

Table I-A: Use case description of arrange appointment use case

Brief Description	Patient arranges appointment with Physician using Scheduler
Actors	Patient, Physician
Flow of Events	Patient arranges appointment with physician using scheduler
Alternative Flow	None
Precondition	None
Post condition	Appointment is made

Table I-B: Use Case description of cancel appointment

Brief Description	Patient cancels appointment with Physician using Scheduler
Actors	Patient, Physician
Flow of Events	Patient cancels appointment with physician using scheduler
Alternative Flow	None
Precondition	None
Post condition	Appointment is cancelled

Table I-C: Use Case description medical history

Brief Description	Physician checks medical History to Prescribe Medication
Actors	Patient, physician
Flow of Events	Physician checks patient's Medical History to prescribe Medication
Alternative Flow	Medication is not prescribed
Precondition	None
Post condition	Prescribe medication

Table I-D: Use Case description print medication

Brief Description	Final Medication is Printed
Actors	Patient, computer
Flow of Events	As Cross-Checks for variance among Medication and present Medications/Medical History treatment
Alternative Flow	Medication is not printed

Brief Description	Final Medication is Printed
Precondition	Cross-checks for disagreement among medication and present medications/Medical History prescription
Post condition	None

Table I-E: Use Case description scheduler use case

Brief Description	Arranges and cancels appointment
Actors	Patient, Physician
Flow of Events	If patient asks for an appointment and scheduler arranges it with physician when he/she is free. If patient asks for that scheduler cancels his/her appointment
Alternative Flow	None
Precondition	None
Post condition	None

Table I-F: Use Case description cross check for conflict

Brief Description	Computer checks the Medication and Current Medication/Medical history prescription.
Actors	Computer
Flow of Events	Computer Cross-Checks for Conflict Between Medication and Current Medications/Medical History Prescription .
Alternative Flow	None
Precondition	None
Post condition	Forward to pharmacy

Table I-G: Use Case description forward to pharmacy

Brief Description	If cross-check performed by computer result's is positive than this use is executed.
Actors	Computer
Flow of Events	If cross-check performed by computer result's is positive than patient is forwarded to pharmacy.
Alternative Flow	None
Precondition	None
Post condition	Medication is provided

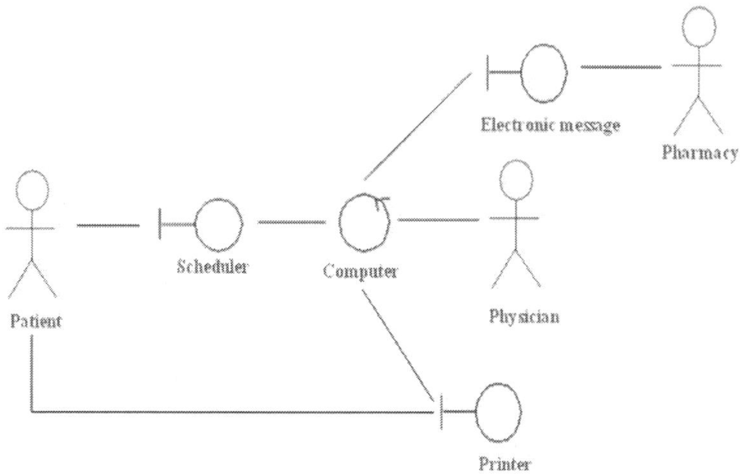

Fig. I-B: Analysis model for hospital and medical sector

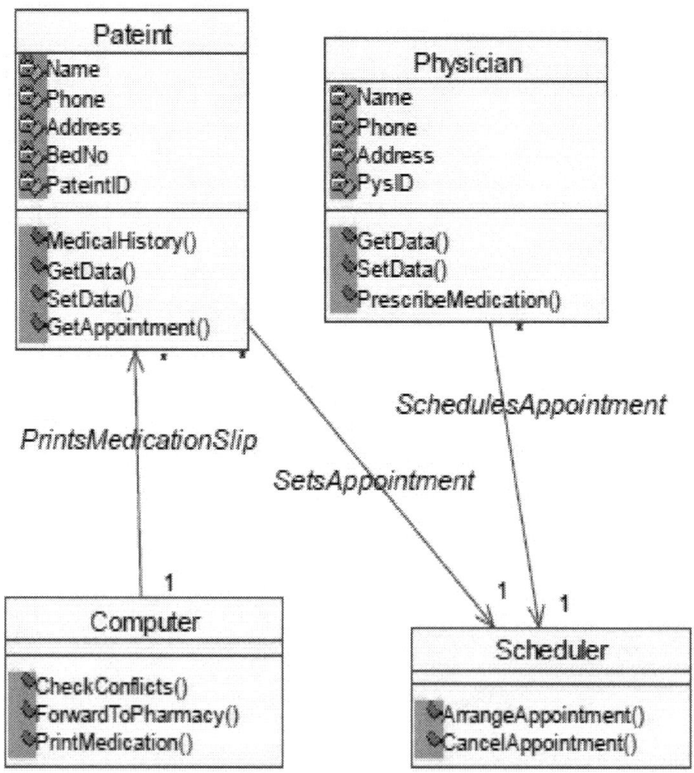

Fig. I-C: Class Diagram of hospital and medical sector

Code of the Example

```
#include "Computer.h"
//##ModelId=4F7ACECA0126
Computer::CheckConflicts()
{
}
//##ModelId=4F7ACED300D6
Computer::ForwardToPharmacy()
{
}
//##ModelId=4F7ACEE1013A
Computer::PrintMedication()
{
}
#include "Pateint.h"
//##ModelId=4F7ACE5D0112
Pateint::MedicalHistory()
{
}
//##ModelId=4F7ACE6600CC
Pateint::GetData()
{
}
//##ModelId=4F7ACE790112
Pateint::SetData()
{
}
//##ModelId=4F7ACE7C01F8
Pateint::GetAppointment()
{
}

#include "Physician.h"
//##ModelId=4F7ACE070036
Physician::GetData()
{
}
```

```
//##ModelId=4F7ACE0E0086
Physician::SetData()
{
}
//##ModelId=4F7ACE1503CE
Physician::PrescribeMedication()
{
}
#include "Scheduler.h"
//##ModelId=4F7ACE910036
Scheduler::ArrangeAppointment()
{
}
//##ModelId=4F7ACE9803B0
Scheduler::CancelAppointment()
{
}
```

Made in the USA
Las Vegas, NV
19 September 2026

4129486f-2fb2-41e2-a1a8-a127670d7f26R01